Selected Titles in This Series

729 Michael Grosser, Eva Farkas, Michael Kunzinger, and Roland Steinbauer, On the foundations of nonlinear generalized functions I and II, 2001

728 Laura Smithies, Equivariant analytic localization of group representations, 2001

727 Anthony D. Blaom, A geometric setting for Hamiltonian perturbation theory, 2001

726 Victor L. Shapiro, Singular quasilinearity and higher eigenvalues, 2001

725 Jean-Pierre Rosay and Edgar Lee Stout, Strong boundary values, analytic functionals, and nonlinear Paley-Wiener theory, 2001

724 Lisa Carbone, Non-uniform lattices on uniform trees, 2001

723 Deborah M. King and John B. Strantzen, Maximum entropy of cycles of even period, 2001

722 Hernán Cendra, Jerrold E. Marsden, and Tudor S. Ratiu, Lagrangian reduction by stages, 2001

721 Ingrid C. Bauer, Surfaces with $K^2 = 7$ and $p_g = 4$, 2001

720 Palle E. T. Jorgensen, Ruelle operators: Functions which are harmonic with respect to a transfer operator, 2001

719 Steve Hofmann and John L. Lewis, The Dirichlet problem for parabolic operators with singular drift terms, 2001

718 Bernhard Lani-Wayda, Wandering solutions of delay equations with sine-like feedback, 2001

717 Ron Brown, Frobenius groups and classical maximal orders, 2001

716 John H. Palmieri, Stable homotopy over the Steenrod algebra, 2001

715 W. N. Everitt and L. Markus, Multi-interval linear ordinary boundary value problems and complex symplectic algebra, 2001

714 Earl Berkson, Jean Bourgain, and Aleksander Pełczynski, Canonical Sobolev projections of weak type $(1,1)$, 2001

713 Dorina Mitrea, Marius Mitrea, and Michael Taylor, Layer potentials, the Hodge Laplacian, and global boundary problems in nonsmooth Riemannian manifolds, 2001

712 Raúl E. Curto and Woo Young Lee, Joint hyponormality of Toeplitz pairs, 2001

711 V. G. Kac, C. Martinez, and E. Zelmanov, Graded simple Jordan superalgebras of growth one, 2001

710 Brian Marcus and Selim Tuncel, Resolving Markov chains onto Bernoulli shifts via positive polynomials, 2001

709 B. V. Rajarama Bhat, Cocylces of CCR flows, 2001

708 William M. Kantor and Ákos Seress, Black box classical groups, 2001

707 Henning Krause, The spectrum of a module category, 2001

706 Jonathan Brundan, Richard Dipper, and Alexander Kleshchev, Quantum Linear groups and representations of $GL_n(\mathbb{F}_q)$, 2001

705 I. Moerdijk and J. J. C. Vermeulen, Proper maps of toposes, 2000

704 Jeff Hooper, Victor Snaith, and Min van Tran, The second Chinburg conjecture for quaternion fields, 2000

703 Erik Guentner, Nigel Higson, and Jody Trout, Equivariant E-theory for C^*-algebras, 2000

702 Ilijas Farah, Analytic quotients: Theory of liftings for quotients over analytic ideals on the integers, 2000

701 Paul Selick and Jie Wu, On natural coalgebra decompositions of tensor algebras and loop suspensions, 2000

700 Vicente Cortés, A new construction of homogeneous quaternionic manifolds and related geometric structures, 2000

(*Continued in the back of this publication*)

On the Foundations of Nonlinear Generalized Functions I and II

of the
American Mathematical Society

Number 729

On the Foundations of Nonlinear Generalized Functions I and II

Michael Grosser
Eva Farkas
Michael Kunzinger
Roland Steinbauer

September 2001 • Volume 153 • Number 729 (end of volume) • ISSN 0065-9266

American Mathematical Society
Providence, Rhode Island

2000 *Mathematics Subject Classification.*
Primary 46F30; Secondary 26E15, 46E50, 35D05.

Library of Congress Cataloging-in-Publication Data
On the foundations of nonlinear generalized functions I and II / Michael Grosser... [et al.].
    p. cm. — (Memoirs of the American Mathematical Society, ISSN 0065-9266 ; no. 729)
"September 2001, volume 153, number 729 (end of volume)."
Includes bibliographical references.
ISBN 0-8218-2729-4
    1. Theory of distributions (Functional analysis). I. Grosser, Michael. II. Series.
QA3.A57 no. 729
[QA324]
510 s—dc21
[515′.782]   2001032795

Memoirs of the American Mathematical Society

This journal is devoted entirely to research in pure and applied mathematics.

Subscription information. The 2001 subscription begins with volume 149 and consists of six mailings, each containing one or more numbers. Subscription prices for 2001 are $494 list, $395 institutional member. A late charge of 10% of the subscription price will be imposed on orders received from nonmembers after January 1 of the subscription year. Subscribers outside the United States and India must pay a postage surcharge of $31; subscribers in India must pay a postage surcharge of $43. Expedited delivery to destinations in North America $35; elsewhere $130. Each number may be ordered separately; *please specify number* when ordering an individual number. For prices and titles of recently released numbers, see the New Publications sections of the *Notices of the American Mathematical Society*.

Back number information. For back issues see the *AMS Catalog of Publications*.

Subscriptions and orders should be addressed to the American Mathematical Society, P. O. Box 845904, Boston, MA 02284-5904. *All orders must be accompanied by payment.* Other correspondence should be addressed to Box 6248, Providence, RI 02940-6248.

Copying and reprinting. Individual readers of this publication, and nonprofit libraries acting for them, are permitted to make fair use of the material, such as to copy a chapter for use in teaching or research. Permission is granted to quote brief passages from this publication in reviews, provided the customary acknowledgment of the source is given.

Republication, systematic copying, or multiple reproduction of any material in this publication is permitted only under license from the American Mathematical Society. Requests for such permission should be addressed to the Assistant to the Publisher, American Mathematical Society, P. O. Box 6248, Providence, Rhode Island 02940-6248. Requests can also be made by e-mail to `reprint-permission@ams.org`.

Memoirs of the American Mathematical Society is published bimonthly (each volume consisting usually of more than one number) by the American Mathematical Society at 201 Charles Street, Providence, RI 02904-2294. Periodicals postage paid at Providence, RI. Postmaster: Send address changes to Memoirs, American Mathematical Society, P. O. Box 6248, Providence, RI 02940-6248.

© 2001 by the American Mathematical Society. All rights reserved.
This publication is indexed in *Science Citation Index*®, *SciSearch*®, *Research Alert*®, *CompuMath Citation Index*®, *Current Contents*®/*Physical, Chemical & Earth Sciences*.
Printed in the United States of America.

∞ The paper used in this book is acid-free and falls within the guidelines established to ensure permanence and durability.
Visit the AMS home page at URL: `http://www.ams.org/`

10 9 8 7 6 5 4 3 2 1   06 05 04 03 02 01

Contents

Abstract   ix

Preface   xi

Part 1. On the Foundations of Nonlinear Generalized Functions I   1

Chapter 1. Introduction   3

Chapter 2. Notation and Terminology   6

Chapter 3. Scheme of construction   8

Chapter 4. Calculus   11

Chapter 5. C- and J-formalism   16

Chapter 6. Calculus on $U_\varepsilon(\Omega)$ 22

Chapter 7. Construction of a diffeomorphism invariant Colombeau algebra   27
     7.1. The basis for the definition of the algebra   27
     7.2. The approach taken by J. Jelínek   30
     7.3. Stability under differentiation   32
     7.4. Diffeomorphism invariance   33

Chapter 8. Sheaf properties   39

Chapter 9. Separating the basic definition from testing   41

Chapter 10. Characterization results   43

Chapter 11. Differential Equations   52

Part 2. On the Foundations of Nonlinear Generalized Functions II   55

Chapter 12. Introduction to Part 2   57

Chapter 13. A simple condition equivalent to negligibility   58

Chapter 14. Some more calculus   61

Chapter 15. Non-injectivity of the canonical homomorphism from $\mathcal{G}^d(\Omega)$ into $\mathcal{G}^e(\Omega)$ 64
     15.1. Proof of the estimates (15.4)   67
     15.2. Proof of smoothness of P 70

15.3. Proof of moderateness of P — 70
15.4. Proof of $P \notin \mathcal{N}^d$ — 70
15.5. Proof of $P \in \mathcal{N}^e$ — 71

Chapter 16. Classification of smooth Colombeau algebras between $\mathcal{G}^d(\Omega)$ and $\mathcal{G}^e(\Omega)$ — 74
16.1. The development leading from $\mathcal{G}^e(\Omega)$ to $\mathcal{G}^d(\Omega)$ — 74
16.2. Classification of test objects — 76
16.3. Classification of full smooth Colombeau algebras — 77

Chapter 17. The algebra $\mathcal{G}^2$; classification results — 82

Chapter 18. Concluding remarks — 90

Acknowledgments — 91

Bibliography — 92

Abstract

In part 1 we construct a diffeomorphism invariant (Colombeau-type) differential algebra canonically containing the space of distributions in the sense of L. Schwartz. Employing differential calculus in infinite dimensional (convenient) vector spaces, previous attempts in this direction are unified and completed. Several classification results are achieved and applications to nonlinear differential equations involving singularities are given.

Part 2 gives a comprehensive analysis of algebras of Colombeau-type generalized functions in the range between the diffeomorphism-invariant quotient algebra $\mathcal{G}^d = \mathcal{E}_M/\mathcal{N}$ introduced in part 1 and Colombeau's original algebra $\mathcal{G}^e$. Three main results are established: First, a simple criterion describing membership in $\mathcal{N}$ (applicable to all types of Colombeau algebras) is given. Second, two counterexamples demonstrate that $\mathcal{G}^d$ is not injectively included in $\mathcal{G}^e$. Finally, it is shown that in the range "between" $\mathcal{G}^d$ and $\mathcal{G}^e$ only one more construction leads to a diffeomorphism invariant algebra. In analyzing the latter, several classification results essential for obtaining an intrinsic description of $\mathcal{G}^d$ on manifolds are derived.

2000 Mathematics Subject Classification. Primary 46F30; Secondary 26E15, 46E50, 35D05.

Keywords and phrases. Algebras of generalized functions, Colombeau algebras, calculus on infinite dimensional spaces, convenient vector spaces, diffeomorphism invariance.

Received by the editor January 11, 2000.

Preface

The present memoir consists of the papers *On the foundations of nonlinear generalized functions I* by E. Farkas, M. Grosser, M. Kunzinger and R. Steinbauer and *On the foundations of nonlinear generalized functions II* by M. Grosser. Its purpose is to develop a diffeomorphism invariant theory of algebras of generalized functions that can serve as a basis for a nonlinear geometric theory of generalized functions on differentiable manifolds.

The basic setup we are using to this end is Colombeau's theory of algebras of generalized functions. From the very beginning the main applications of this theory have been in the field of nonlinear partial differential equations in the presence of singularities. Rather recently, applications in mathematical physics, in particular in general relativity have demonstrated the need for a diffeomorphism invariant version of the theory retaining the optimal consistency properties with respect to classical operations on distributions displayed by Colombeau's construction on $\mathbb{R}^n$. Work in this direction was pioneered by J. F. Colombeau and A. Meril in [**13**] and by J. Jelínek in [**26**]. Since some of the main "technical ingredients" of Colombeau algebras (regularization of distributions via convolution, moment conditions, ...) are clearly not invariant under coordinate changes, the development of such a theory requires an in-depth analysis of the fundamental building blocks that lie at the core of Colombeau's construction. This analysis by necessity has to draw from a rather surprising variety of mathematical fields. In particular, calculus in infinite dimensional (convenient) vector spaces plays a decisive role in our approach.

In our presentation we have laid particular emphasis on working out the inner constraints of constructing diffeomorphism invariant Colombeau-type differential algebras (e.g. by introducing a general scheme of construction applicable to all types of Colombeau algebras) and to determine *all* possible choices for such constructions: Part 2 of this monograph gives a complete classification of (full) Colombeau algebras and singles out exactly two versions that in fact meet all the requirements for such a theory.

We hope that this work may serve as a basic reference for the growing number of applications of algebras of generalized functions to geometric questions.

Michael Grosser,
Eva Farkas,
Michael Kunzinger,
Roland Steinbauer

Part 1

On the Foundations of Nonlinear Generalized Functions I

CHAPTER 1

Introduction

In his celebrated impossibility result ([**38**]), L. Schwartz demonstrated that the space $\mathcal{D}'(\Omega)$ of distributions over some open subset Ω of $\mathbb{R}^n$ cannot be embedded into an associative commutative algebra $(\mathcal{A}(\Omega), +, \circ)$ satisfying

(i) $\mathcal{D}'(\Omega)$ is linearly embedded into $\mathcal{A}(\Omega)$ and $f(x) \equiv 1$ is the unity in $\mathcal{A}(\Omega)$.
(ii) There exist derivation operators $\partial_i : \mathcal{A}(\Omega) \to \mathcal{A}(\Omega)$ ($i = 1, \ldots, n$) that are linear and satisfy the Leibnitz rule.
(iii) $\partial_i|_{\mathcal{D}'(\Omega)}$ is the usual partial derivative ($i = 1, \ldots, n$).
(iv) $\circ|_{\mathcal{C}(\Omega) \times \mathcal{C}(\Omega)}$ coincides with the pointwise product of functions.

Since this result remains valid upon replacing $\mathcal{C}(\Omega)$ by $\mathcal{C}^k(\Omega)$ for any finite k, the best possible result would consist in constructing an embedding of $\mathcal{D}'(\Omega)$ as above with (iv) replaced by

(iv') $\circ|_{\mathcal{C}^\infty(\Omega) \times \mathcal{C}^\infty(\Omega)}$ coincides with the pointwise product of functions.

The actual construction of differential algebras satisfying these optimal properties is due to J. F. Colombeau ([**9**], [**10**], [**11**], [**12**]). The need for algebras of this type arises, for example, from the necessity of considering non-linear PDEs where either the respective coefficients, the data or the prospective solutions are non-smooth. Classical linear distribution theory clearly does not permit the treatment of such problems. Colombeau algebras, on the other hand, have proven to be a useful tool for analyzing such questions (for applications in nonlinear PDEs, cf. e.g., [**4**], [**5**], [**16**], [**14**], [**15**], [**27**], [**33**], for applications to numerics, see e.g., [**3**], [**6**], [**7**], for applications in mathematical physics, e.g., [**42**], [**30**], [**23**] as well as the literature cited in these works). For alternative approaches to algebras of generalized functions, cf. [**35**], [**36**].

Since Colombeau's monograph [**9**], there have been introduced a considerable number of variants of Colombeau algebras, many of them adapted to special purposes. From the beginning, however, the question of the functor property of the construction was at hand as a crucial one: If $\mu : \tilde{\Omega} \to \Omega$ denotes a diffeomorphism between open subsets $\tilde{\Omega}, \Omega$ of $\mathbb{R}^s$, is it possible to extend the operation $\mu^* : f \mapsto f \circ \mu$ on smooth distributions on Ω to an operation $\hat{\mu}$ on the Colombeau algebra such that $(\mu \circ \nu)\hat{\ } = \hat{\nu} \circ \hat{\mu}$ and $(\mathrm{id})\hat{\ } = \mathrm{id}$ are satisfied? To phrase it differently, is it possible to achieve a diffeomorphism invariant construction of Colombeau algebras? As long as this question could not be answered in the positive, there remained the serious objection that there is no way of defining such algebras on manifolds, based on intrinsic terms, exclusively. (This discussion does not take into account the so called "special" or "simplified" variant of Colombeau's algebra whose elements are classes of nets of smooth functions indexed by $\varepsilon > 0$ (cf. [**33**], p. 109). Although diffeomorphism-invariant, these algebras lack a canonical embedding of distributions ([**17**], [**39**]), so we do not consider them here.)

The first variants of Colombeau algebras, though serving as a valuable tool in the treatment of non-linear problems, indeed did not have the property of diffeomorphism invariance: Some of the key ingredients used in defining them (in particular, the "test objects" (see chapter 3) being employed as well as the definition of the subsets $\mathcal{A}_q(\mathbb{R}^s)$ of the set of all test objects) turned out not to be invariant under the natural action of a diffeomorphism.

Colombeau and Meril in their paper [13] made the first decisive steps to remove this flaw by proposing a construction of Colombeau algebras which they claimed to be diffeomorphism invariant. As an essential tool, they had to use calculus on locally convex spaces. However, they did not give the details of the application of that calculus; moreover, their definition of the objects constituting the Colombeau algebra was not unambiguous and, which amounts to the most serious objection, their notion of test objects still was not preserved under the action of the diffeomorphism. Nevertheless, despite these defects (which, apparently, went unnoticed by nearly all workers in the field) their construction was quoted and used many times (see, e.g., [29], [18], [1], [43], [40], [32], [41], [19], [42], [2], [30]). It was only in 1998 that J. Jelínek in [26] pointed out the error in [13] by giving a (rather simple) counterexample. In the same paper, he presented another version of the theory avoiding the shortcomings of [13] and forming the basis for the approach taken here.

Part 1 of this memoir is organized as follows: After fixing notation and terminology in chapter 2, a general scheme of construction for diffeomorphism-invariant Colombeau-type algebras of generalized functions is introduced in chapter 3. Chapter 4 gives a quick overview of calculus in convenient vector spaces providing the necessary results for the development of the theory. Especially with a view to applications (in particular: partial differential equations) we feel that this approach has several advantages over the concept of Silva-differentiability employed so far. Chapter 5 introduces a translation formalism that allows to freely switch between what we call the C- and J- (*Colombeau-* and *Jelinek-*) formalism of the diffeomorphism invariant theory to be constructed in chapter 7. In the actual construction of this algebra, smooth functions defined on sets denoted by $U_\varepsilon(\Omega)$ play a central rôle. Differentials of such functions are of utmost importance in the development of the theory. However, $U_\varepsilon(\Omega)$ is not a linear space. Chapter 6 thus provides the framework necessary for doing calculus on $U_\varepsilon(\Omega)$. A complete presentation of the resulting diffeomorphism invariant algebra, based on the general construction scheme of chapter 3, is the focus of chapter 7. The sheaf-theoretic properties of this algebra are discussed in chapter 8. This is followed by a short chapter on the separation of testing procedures and definition of objects in algebras of generalized functions. Chapter 10 provides several new characterizations of the fundamental building blocks $\mathcal{E}_M$ and $\mathcal{N}$ of the algebra. In particular, these characterizations will constitute the key ingredient in obtaining an intrinsic description of the theory on manifolds ([24]). Finally, we present some applications to partial differential equations in chapter 11.

Part 2 of the present work gives a comprehensive analysis of algebras of Colombeau-type generalized functions in the range between the diffeomorphism-invariant quotient algebra $\mathcal{G}^d = \mathcal{E}_M/\mathcal{N}$ introduced in chapter 7 and (the smooth version of) Colombeau's original algebra $\mathcal{G}^e$ introduced in [10] (which, to be sure, is the standard version among those being independent of the choice of a particular approximation of the delta distribution). Three main results are established: First, a simple criterion describing membership in $\mathcal{N}$ (applicable to all types of Colombeau algebras) is given (chapter 13). Second, two counterexamples demonstrate that $\mathcal{G}^d$ is not injectively included in $\mathcal{G}^e$ (chapter 15); their construction is based on a completeness theorem for spaces of smooth functions in the sense of chapters 4 and 6 (chapter 14). Finally, it is shown that in the range "between" $\mathcal{G}^d$ and $\mathcal{G}^e$ only one more construction leads to a diffeomorphism invariant algebra. In analyzing the latter, several classification results essential for obtaining an intrinsic description of $\mathcal{G}^d$ on manifolds are derived (chapters 16, 17). The concluding chapter 18 points out that also weaker invariance properties than with respect to all diffeomorphisms should be envisaged for Colombeau algebras, in particular regarding applications.

CHAPTER 2

Notation and Terminology

Throughout this work, Ω, $\tilde{\Omega}$ will denote non-empty open subsets of $\mathbb{R}^s$. For any $A \subseteq \mathbb{R}^s$, A° denotes its interior. $\mathcal{C}^\infty(\Omega)$ is the space of smooth, complex valued functions on Ω. If $f \in \mathcal{C}^\infty(\Omega)$ then Df denotes its (total) derivative. Also, we set $\check{f}(x) = f(-x)$. On any cartesian product, pr_i denotes the projection onto the i-th factor. For $r \in \mathbb{R}$, $[r]$ is the largest integer $\leq r$. We set $I = (0, 1]$. Concerning locally convex spaces our basic reference is [37]. In particular, by a locally convex space we mean a vector space endowed with a locally convex Hausdorff topology. The space of test functions (i.e., compactly supported smooth functions) on Ω is denoted by $\mathcal{D}(\Omega)$ and is equipped with its natural (LF)-topology; its dual, the space of distributions on Ω is termed $\mathcal{D}'(\Omega)$. The action of any $u \in \mathcal{D}'(\Omega)$ on a test function φ will be written as $\langle u, \varphi \rangle$. δ denotes the Dirac delta distribution. $K \subset\subset A$ ($A \subseteq \mathbb{R}^s$) means that K is a compact subset of A°. For $K \subset\subset \Omega$, $\mathcal{D}_K(\Omega)$ is the space of smooth functions on Ω supported in K. We set

$$\mathcal{A}_0(\Omega) = \{\varphi \in \mathcal{D}(\Omega) \mid \int \varphi(\xi)\, d\xi = 1\}$$

$$\mathcal{A}_q(\Omega) = \{\varphi \in \mathcal{A}_0(\Omega) \mid \int \xi^\alpha \varphi(\xi)\, d\xi = 0,\ 1 \leq |\alpha| \leq q,\ \alpha \in \mathbb{N}_0^s\} \qquad (q \in \mathbb{N})$$

$\mathcal{A}_{q0}(\Omega)$ is the linear subspace of $\mathcal{D}(\Omega)$ parallel to the affine space $\mathcal{A}_q(\Omega)$ ($q \in \mathbb{N}_0$). For any maps f, g, h such that $g \circ f$ and $f \circ h$ are defined we set $g_*(f) := g \circ f$ and $h^*(f) := f \circ h$. ∂_i resp. ∂^α always stand for $\frac{\partial}{\partial x_i}$ resp. $\frac{\partial^{|\alpha|}}{\partial x^\alpha}$ ($\alpha \in \mathbb{N}_0^s$).

For any locally convex space F the space $\mathcal{C}^\infty(\Omega, F)$ of smooth functions from Ω into F will always carry the topology of uniform convergence in all derivatives on compact subsets of Ω. In particular, a subset $\mathcal{B}$ of this space will be said to be bounded if, for any $K \subset\subset \Omega$ and any $\alpha \in \mathbb{N}_0^s$, the set $\{\partial^\alpha \phi(x) \mid \phi \in \mathcal{B}, x \in K\}$ is bounded in F. Observe that in case that the image of a map $\phi \in \mathcal{C}^\infty(\Omega, F)$ is contained in some affine subspace F_0 of F then the derivatives of ϕ take their values in the linear subspace parallel to F_0. For locally convex spaces E, F the space $\mathcal{C}^\infty(E, F)$ (resp. $\mathcal{C}^\infty(E)$ for $F = \mathbb{C}$) is introduced in chapter 4.

In what follows, $\mathcal{A}_0(\mathbb{R}^s)$ may be replaced by any closed affine subspace of $\mathcal{D}(\mathbb{R}^s)$. By $\mathcal{C}_b^{[\infty,\Omega]}(I \times \Omega, \mathcal{A}_0(\mathbb{R}^s))$ we denote the space of all maps $\phi : I \times \Omega \to \mathcal{A}_0(\mathbb{R}^s)$ which are smooth with respect to the second argument and bounded in the sense that the corresponding map $\hat{\phi} : I \to \mathcal{C}^\infty(\Omega, \mathcal{A}_0(\mathbb{R}^s))$ has a bounded image as defined above, i.e., for every $K \subset\subset \Omega$ and any $\alpha \in \mathbb{N}_0^s$, the set $\{\partial^\alpha(\hat{\phi}(\varepsilon))(x) \mid \varepsilon \in I, x \in K\}$ is bounded in $\mathcal{A}_0(\mathbb{R}^s)$ resp. $\mathcal{D}(\mathbb{R}^s)$, which, in turn, is equivalent to saying that

1. for every K as above and any $\alpha \in \mathbb{N}_0^s$, the supports of all $\partial_x^\alpha \phi(\varepsilon, x)$ ($\varepsilon \in I$, $x \in K$) are contained in some fixed bounded set (depending only on K) and

2. $\sup\{|\partial_\xi^\beta(\partial_x^\alpha(\hat\phi(\varepsilon))(x))(\xi)|\varepsilon\in I,\ x\in K,\ \xi\in\mathbb{R}^s\}$ (or, expressed in terms of ϕ itself) $\sup\{|\partial_\xi^\beta\partial_x^\alpha(\phi(\varepsilon,x))(\xi)|\varepsilon\in I,\ x\in K,\ \xi\in\mathbb{R}^s\}$ is finite.

$\mathcal{C}_b^\infty(I\times\Omega,\mathcal{A}_0(\mathbb{R}^s))$ is the subspace of $\mathcal{C}_b^{[\infty,\Omega]}(I\times\Omega,\mathcal{A}_0(\mathbb{R}^s))$ whose elements are smooth in both arguments. Finally, for any $K\subset\subset\Omega$ and any $q\geq 1$ an element ϕ of $\mathcal{C}_b^\infty(I\times\Omega,\mathcal{D}(\mathbb{R}^s))$ is said to have asymptotically vanishing moments of order q on K if

$$\sup_{x\in K}|\int\xi^\alpha\phi(\varepsilon,x)(\xi)\,d\xi|=O(\varepsilon^q)\quad(1\leq|\alpha|\leq q).$$

For this notion to make sense it is obviously sufficient for ϕ to be defined on $(0,\varepsilon_0]\times K$ for some $\varepsilon_0>0$.

CHAPTER 3

Scheme of construction

As was already pointed out in chapter 1, due to the lack of a canonical embedding of the space of distributions into "special" variants of Colombeau algebras we shall not consider these. Instead, we focus on "full" algebras (in the sense of [29], p. 31), distinguished by the fact that such a canonical embedding is always available. Elements of full Colombeau algebras are equivalence classes of functions R taking as arguments certain pairs (φ, x) consisting of a suitable test function $\varphi \in \mathcal{D}(\mathbb{R}^s)$ and a point x of Ω.

Every (full) Colombeau algebra is constructed according to the following blueprint (where (Di), (Tj) stand for Definition i and Theorem j, respectively). (D5) and (T6)–(T8) are only relevant if a diffeomorphism invariant type of algebra is to be obtained.

(D1) $\mathcal{E}(\Omega)$ (the "basic space", see the remarks below); maps $\sigma : \mathcal{C}^\infty(\Omega) \to \mathcal{E}(\Omega)$, $\iota : \mathcal{D}'(\Omega) \to \mathcal{E}(\Omega)$.

(D2) Derivations D_i on $\mathcal{E}(\Omega)$ ($i = 1, \ldots, s$) extending the operators $\frac{\partial}{\partial x_i}$ of partial differentiation on $\mathcal{D}'(\Omega)$ resp. on $\mathcal{C}^\infty(\Omega)$, i.e., $D_i \circ \iota = \iota \circ \frac{\partial}{\partial x_i}$ and $D_i \circ \sigma = \sigma \circ \frac{\partial}{\partial x_i}$.

(D3) $\mathcal{E}_M(\Omega)$ ($\subseteq \mathcal{E}(\Omega)$; the subspace of "moderate" functions).

(D4) $\mathcal{N}(\Omega)$ ($\subseteq \mathcal{E}(\Omega)$; the subspace of "negligible" functions).

(T1) $\iota(\mathcal{D}'(\Omega)) \subseteq \mathcal{E}_M(\Omega), \sigma(\mathcal{C}^\infty(\Omega)) \subseteq \mathcal{E}_M(\Omega), (\iota - \sigma)(\mathcal{C}^\infty(\Omega)) \subseteq \mathcal{N}(\Omega)$; $\iota(\mathcal{D}'(\Omega)) \cap \mathcal{N}(\Omega) = \{0\}$.

(T2) $\mathcal{E}_M(\Omega)$ is a subalgebra of $\mathcal{E}(\Omega)$.

(T3) $\mathcal{N}(\Omega)$ is an ideal in $\mathcal{E}_M(\Omega)$.

(T4) $\mathcal{E}_M(\Omega)$ is invariant under each D_i.

(T5) $\mathcal{N}(\Omega)$ is invariant under each D_i.

(D5) For each diffeomorphism $\mu : \tilde{\Omega} \to \Omega$, a map $\bar{\mu} : D_{\tilde{\Omega}} \to D_\Omega$ [1] is defined in a functorial way such that its "transpose" $\hat{\mu} : \mathcal{E}(\Omega) \to \mathcal{E}(\tilde{\Omega})$, $\hat{\mu}(R) := R \circ \bar{\mu}$, extends the usual effect μ has on distributions, i.e., $\hat{\mu} \circ \iota = \iota \circ \mu^*$ where for $u \in \mathcal{D}'(\Omega)$, $\mu^* u$ is defined by $\langle \mu^* u, \varphi \rangle := \langle u, (\varphi \circ \mu^{-1}) \cdot |\det D\mu^{-1}| \rangle$. Similarly, we require $\hat{\mu} \circ \sigma = \sigma \circ \mu^*$ on $\mathcal{C}^\infty(\Omega)$.

(T6) The class of "scaled test objects" (see below) is invariant under the action induced by μ.

(T7) $\mathcal{E}_M$ is invariant under $\hat{\mu}$, i.e., $\hat{\mu}$ maps $\mathcal{E}_M(\Omega)$ into $\mathcal{E}_M(\tilde{\Omega})$.

(T8) $\mathcal{N}$ is invariant under $\hat{\mu}$, i.e., $\hat{\mu}$ maps $\mathcal{N}(\Omega)$ into $\mathcal{N}(\tilde{\Omega})$.

(D6) $\mathcal{G}(\Omega) := \mathcal{E}_M(\Omega)/\mathcal{N}(\Omega)$.

---

[1] Concerning $D_{\tilde{\Omega}}, D_\Omega$, see the remark on (D1) below.

3. SCHEME OF CONSTRUCTION

For $R \in \mathcal{E}_M(\Omega)$, the class $R + \mathcal{N}(\Omega)$ of R in $\mathcal{G}(\Omega)$ will be denoted by $[R]$.

The following comments are intended to motivate and clarify the preceding—admittedly very formal—definition schemes and theorems.

ad (D1): Here, $\mathcal{E}(\Omega)$ denotes some algebra of complex-valued functions having appropriate smoothness properties on a suitable domain $D_\Omega \subseteq \mathcal{D}(\mathbb{R}^s) \times \Omega$. σ has to be an injective algebra homomorphism, whereas ι just has to be linear and injective.

ad (D3), (D4): Membership of $R \in \mathcal{E}(\Omega)$ in $\mathcal{N}(\Omega)$ respectively $\mathcal{E}_M(\Omega)$ depends on the "asymptotic" behaviour of R on certain paths in $\mathcal{D}(\mathbb{R}^s) \times \Omega$, where the second component is constant whereas the first component, depending on ε as parameter, tends to the delta distribution weakly as $\varepsilon \to 0$. Essentially, these paths are obtained by applying the scaling operator $S_\varepsilon : \varphi \mapsto \frac{1}{\varepsilon^s}\varphi(\frac{\cdot}{\varepsilon})$ (thereby introducing the parameter ε) to so-called test objects. Typically, a test object is some fixed element $\varphi \in \mathcal{D}(\mathbb{R}^s)$ satisfying $\int \varphi = 1$ or a suitable bounded family $\phi(\varepsilon, x) \in \mathcal{D}(\Omega)$, parametrized by $\varepsilon \in I$, $x \in \Omega$, where again $\int \phi(\varepsilon, x)(\xi)\, d\xi \equiv 1$. Roughly speaking, R is defined to be negligible if the values R attains on those "scaled test objects" tend to zero faster than any positive power of ε, while it is called moderate if these values are bounded by some fixed (negative) power of ε. In both cases, convergence in each derivative, uniformly on compact subsets of Ω, is required. We will refer to those defining procedures as **testing for negligibility** resp. **moderateness** (see also chapter 9).

ad (T1), (T3): $\mathcal{N}(\Omega)$ has to be large enough to contain all $\sigma(f) - \iota(f)$ ($f \in \mathcal{C}^\infty(\Omega)$) (this renders $\iota\,|_{\mathcal{C}^\infty(\Omega)}$ an algebra homomorphism by passing to a quotient by $\mathcal{N}(\Omega)$), however small enough to intersect $\mathcal{D}'(\Omega)$ just in $\{0\}$ (this guarantees $\mathcal{D}'(\Omega)$ to be contained injectively in the quotient by $\mathcal{N}(\Omega)$). $\mathcal{E}_M(\Omega)$, on the other hand, clearly has to be large enough to contain $\mathcal{C}^\infty(\Omega)$ and $\mathcal{D}'(\Omega)$ (via σ resp. ι), yet small enough such that $\mathcal{N}(\Omega)$ is an ideal in it: This will allow us to form the quotient $\mathcal{E}_M(\Omega)/\mathcal{N}(\Omega)$.

ad (D5): μ^* as defined above extends $\mu^* : f \mapsto f \circ \mu$ where the latter is viewed as the action induced by μ on the smooth distribution $f \in \mathcal{C}^\infty(\Omega)$. Hence we regard distributions (and, in the sequel, non-linear generalized functions) as generalizations of *functions* on the respective open set, acting as functionals on (smooth, compactly supported) densities. This is in agreement with, for example, [25], however has to be distinguished clearly from constructing distributions as distributional *densities*, acting on (smooth, compactly supported) functions, as it is done, e.g., in [20].

ad (T7), (T8): Because of the forms of **(D3)** and **(D4)** as tests to be performed on the elements R of $\mathcal{E}(\Omega)$, with the appropriate type of (scaled) test objects being inserted, **(T7)** as well as **(T8)** follow immediately from **(T6)**, taking into account **(D5)**.

ad (D6): By this definition, $\mathcal{G}(\Omega)$ is a differential algebra containing $\mathcal{D}'(\Omega)$ via ι followed by the canonical quotient map (**(T1)–(T5)**); by abuse of notation, we will denote this embedding also by ι. Each diffeomorphism $\mu : \tilde\Omega \to \Omega$ induces a map $\hat\mu : \mathcal{G}(\Omega) \to \mathcal{G}(\tilde\Omega)$ extending the usual action of μ on distributions such that composition and identities are preserved by $\mu \mapsto \hat\mu$ (**(T7), (T8)**).

Without the requirement of diffeomorphism invariance (as, for example, in [9]), the smoothness property of R mentioned above only needs to refer to the variable x in the pair (φ, x), thus involving only classical calculus. However, as mentioned already in the introduction, to obtain a diffeomorphism invariant algebra we also have to consider smoothness with respect to the test function φ. Therefore, in the following chapter, we are going to outline the elements of calculus on locally convex spaces which are required for the subsequent constructions. The path we will pursue in this respect is different from the approaches taken so far and, in our view, has some decisive advantages over these.

CHAPTER 4

Calculus

In the first versions of Colombeau algebras (on $\mathbb{R}^n$ or open subsets thereof), the main ingredient was the algebra of smooth functions $\varphi \mapsto R(\varphi)$ on the ((LF)-) space $\mathcal{D}$ of test functions (see [9]). Thus, from the very beginning, there had to be a theory of differentiation on (certain non-Banach) locally convex spaces at the basis of the construction of these algebras.

Colombeau's approach in [9] employs the notion of *Silva-differentiability* ([44], [8]) where a map $f : E \supseteq U \to F$ from an open subset U of a locally convex space E into another locally convex space F is called Silva-differentiable in $x \in U$ if there exists a bounded linear map (called $f'(x)$) $E \to F$ such that the restriction of the corresponding remainder function to sufficiently small homothetic images of bounded subsets may be viewed as a map between suitable normed spaces and satisfies a condition thereon which is completely analogous to the classical remainder condition for Fréchet-differentiable maps.

In later versions, Colombeau managed to circumvent this necessity by introducing an additional variable $x \in \mathbb{R}^n$ into R which could carry the burden of smoothness: For the construction of the algebra $\mathcal{G}(\Omega)$ of [10] he now used functions $R(\varphi, x)$ which, for each fixed φ from (a certain affine subspace of) $\mathcal{D}$, are smooth in x (in the usual elementary sense—hence the title of [10]) whereas the dependence on φ is completely arbitrary; φ just plays the rôle of a parameter in this setting. Apart from simplifying the general setup of the theory the introduction of x as a separate variable was also crucial for solving differential equations in $\mathcal{G}(\Omega)$.

However, when Colombeau and Meril in [13] began to develop a diffeomorphism invariant version of the algebra $\mathcal{G}(\Omega)$ of [10], they had to reintroduce the smooth dependence of R on φ: Under the action of a diffeomorphism μ, φ changes to some $\tilde\varphi_x$ depending on x. For the smoothness of the μ-transform of R (which, according to (**D5**), is of the form $(\hat\mu R)(\varphi, x) = R(\bar\mu(\varphi, x)) = R(\tilde\varphi_x, \mu x))$ with respect to x, obviously the smooth dependence of R also on its first argument φ is needed ([13], p. 263). Concerning calculus on locally convex spaces, the authors—as the first of them did already in [9]—refer to [8]. Omitting any details in this respect, they rather invite the reader to admit the respective smoothness properties (p. 263).

Jelínek in [26] includes a section on calculus (items 9–16): In addition to [8], he quotes [44] as reference for some results needed. The relevant statements are formulated in terms of higher Fréchet differentials.

Contrary to the above, we prefer to base our presentation on the notion of smoothness as it is outlined in [28]. This approach seems to us to have a number of striking advantages: On the one hand, the basic definition is very simple and easy to work with, a smooth map between locally convex spaces E, F being one that takes smooth curves $\mathbb{R} \to E$ to smooth curves $\mathbb{R} \to F$ (by composition); the notion of a smooth curve into a locally convex space obviously is without problems. We will denote by $\mathcal{C}^\infty(E,F)$ the space of smooth maps between E and F. For $\mathcal{C}^\infty(E,\mathbb{C})$, we will simply write $\mathcal{C}^\infty(E)$. On the other hand, all the basic theorems of differential calculus can be reconstructed in this setting (see, e.g., the version of the mean value theorem given in 4.5) and more than that (see, e.g., the exponential laws stated in 4.2 and 4.3 below and the differentiable uniform boundedness principle 4.7). As smooth curves are continuous, the above definition of smoothness carries over to open subsets of locally convex spaces. We will make use of this in the sequel and want to note that in any of the theorems of this chapter, we may replace the respective locally convex (domain) spaces by open subsets thereof whenever their linear structure is not needed.

This notion of smoothness is a weaker one than Silva-differentiability but turns out to be equivalent for a huge class of spaces, e.g. those which are complete and Montel so that the two notions coincide in particular on the regular[1] strict inductive limit $\mathcal{D}(\Omega)$ of Fréchet spaces and each closed subspace thereof.

The seeming drawback of this (and any other reasonable such as Colombeau's above-mentioned) theory of differentiation is the fact that smooth maps (resp. their differentials) need no longer be continuous. The fundamental rôle played by continuity in the classical context is taken over by the notion of boundedness: Indeed, the difference quotients of smooth curves converge in a stronger sense than the topological one, so that continuity is not a necessary property for a map to be smooth. In order to be able to test smoothness by composition with suitable families of linear functionals (see, e.g., 4.7) one needs, in addition, a completeness property which is weaker than completeness of the locally convex topology. Separated bornological locally convex spaces which have this property are called *convenient spaces* and are in some sense the most general class of linear spaces in which one can perform differentiation and integration. As for each locally convex space there exists a finer bornological locally convex topology with the same bornology, i.e., the same system of bounded sets, bornologicity of the topology is not essential. It will be enough for our purpose to confine ourselves to the particular case of complete locally convex spaces.

In the sequel, we will endow the space $\mathcal{C}^\infty(\mathbb{R},F)$ of smooth curves into the locally convex space F with the locally convex topology of uniform convergence on compact intervals in each derivative separately. More generally, we may consider on the space $\mathcal{C}^\infty(E,F)$ the initial locally convex topology induced by the pullbacks along smooth curves $\mathbb{R} \to E$. It can be shown that the bounded sets associated with this topology are the same as the ones associated with the topology of uniform convergence on compact subsets in each differential (as defined for such maps in 4.4) separately. Moreover, as mentioned in the introduction, for complete F, the latter is again complete; see chapter 14 for details.

Testing of smoothness is particularly simple in the case of a linear map: A linear map is smooth if and only if it is bounded. $L(E,F)$ will stand for the space of bounded (smooth) linear maps between E, F.

THEOREM 4.1. *A map f from $\mathcal{D}(\Omega)$ into a locally convex space E is smooth if and only if for each $K \subset\subset \Omega$, the restriction of f to $\mathcal{D}_K(\Omega)$ is smooth.*

---

[1]A strict inductive limit $\varinjlim E_\alpha$ is called regular if each bounded subset is contained in some E_α. Note that every strict inductive limit of an increasing sequence E_n is regular, as is $\mathcal{D}(M)$ for any paracompact (not necessarily separable) smooth manifold M.

PROOF. For the non-trivial part of the proof, consider a smooth curve $c : \mathbb{R} \to \mathcal{D}(\Omega)$. Its restriction to any bounded interval J has a relatively compact, hence bounded image. Therefore, c maps J into some $\mathcal{D}_K(\Omega)$ and the same holds for each derivative of c since $\mathcal{D}_K(\Omega)$ is a closed subspace of $\mathcal{D}$. By assumption, $f \circ c$ is smooth on J. Since smoothness is a local property, we are done. □

The obvious generalization of the preceding theorem is true for any strict inductive limit of a sequence of Fréchet spaces. Its trivial part has an important consequence: $\mathcal{D}_K(\Omega)$ being a Fréchet space, the restriction to $\mathcal{D}_K(\Omega)$ of any smooth map f from $\mathcal{D}(\Omega)$ to any metrizable locally convex space E is continuous: Both on $\mathcal{D}_K(\Omega)$ and E the so-called c^∞-topology (see [**28**]) coincides with the metric topology ([**28**], 4.11.(1)); moreover, smooth maps are continuous with respect to the c^∞-topology ([**28**], p. 8).

One of the particular features of the Frölicher-Kriegl-theory which considerably simplify its application is the *exponential law* (cf. Theorem 3.12 and Corollary 3.13 in [**28**]):

THEOREM 4.2. *Let E, F, G be locally convex spaces. Then the two spaces $\mathcal{C}^\infty(E \times F, G)$ and $\mathcal{C}^\infty(E, \mathcal{C}^\infty(F, G))$ are isomorphic algebraically and bornologically, i.e., they have the same bounded sets.*

Replacing $\mathcal{C}^\infty$ by L in 4.2 yields the exponential law for linear smooth maps. By iteration one obtains (see Proposition 5.2 in [**22**]):

THEOREM 4.3. *Let $n, k \in \mathbb{N}$ and E_i, F $(i = 1, \ldots, n+k)$ locally convex spaces. Then there is a bornological isomorphism*

$$L(E_1, \ldots, E_{n+k}; F) \cong L(E_1, \ldots, E_n; L(E_{n+1}, \ldots, E_{n+k}; F)).$$

For later use, we present the analoga of items 10–16 in [**26**] in the setting of [**28**]:

THEOREM 4.4. *(Theorem 3.18 and Corollary 5.11 in [**28**]) Let E, F be locally convex spaces. Then the differentiation operator $\mathrm{d} : \mathcal{C}^\infty(E, F) \to \mathcal{C}^\infty(E, L(E, F))$ given by*

$$\mathrm{d}f(x)v := \lim_{t \to 0} \frac{f(x+tv) - f(x)}{t}$$

exists and is linear and bounded (smooth). Hence, for $n \in \mathbb{N}$ one can form the iterated differentiation operator

$$\mathrm{d}^n : \mathcal{C}^\infty(E, F) \to \mathcal{C}^\infty(E, L(E, \ldots, L(E; F) \ldots)) \cong \mathcal{C}^\infty(E, L(E, \ldots, E; F))$$

which is smooth and linear and has values in $\mathcal{C}^\infty(E, L_{sym}(E, \ldots, E; F))$, where $L_{sym}(E, \ldots, E; F)$ stands for the space of smooth n-linear symmetric maps between $E \times \cdots \times E$ and F. Also, the chain rule holds:

$$\mathrm{d}(f \circ g)(x)v = \mathrm{d}f(g(x))\mathrm{d}g(x)v.$$

It is shown in [**28**], 1.4, that, given a curve which is smooth from (an open neighborhood of) $\mathbb{R} \supseteq [a, b]$ to E, the difference quotient $\frac{c(b)-c(a)}{b-a}$ is an element of $\overline{\mathrm{conv}}\{c'(t) : t \in [a, b]\}$, where $\overline{\mathrm{conv}}$ denotes the closed convex hull. By virtue of the chain rule given in 4.4, this is equivalent to

PROPOSITION 4.5. *(Mean Value Theorem) Let $f : E \supseteq U \to F$ be smooth, where U is an open neighborhood of a segment $[x, x+v] \subseteq E$. Then*

$$f(x+v) - f(x) \in \overline{\mathrm{conv}}\{\mathrm{d}f(x+tv)(v) : t \in [0,1]\}.$$

As a consequence of 4.2, for each smooth map $f \in \mathcal{C}^\infty(F, G)$, the maps $f_* : \mathcal{C}^\infty(E, F) \to \mathcal{C}^\infty(E, G)$ and $f^* : \mathcal{C}^\infty(G, E) \to \mathcal{C}^\infty(F, E)$ are smooth. In particular, for a smooth map $f \in \mathcal{C}^\infty(E \times F, G)$ we may define smooth linear "operators of partial differentials" d_1, d_2 as

$$d_1 := (\iota_E^*)_* \circ d : \mathcal{C}^\infty(E \times F, G) \to \mathcal{C}^\infty(E \times F, L(E, G))$$

and

$$d_2 := (\iota_F^*)_* \circ d : \mathcal{C}^\infty(E \times F, G) \to \mathcal{C}^\infty(E \times F, L(F, G)),$$

where ι_E, ι_F denote the natural embeddings of E resp. F into $E \times F$. Obviously, we have

$$d_1 f(x)(v) = df(x)(\iota_E(v)) = \lim_{t \to 0} \frac{f(x + t\iota_E(v)) - f(x)}{t},$$

which yields an alternative definition of d_1, which makes sense also for maps $f : E \times F \to G$ which are not a priori known to be smooth on $E \times F$.

PROPOSITION 4.6. *A map on $E \times F$ is smooth if and only if both partial differentials d_1, d_2 exist and are smooth as maps on $E \times F$. In this case the differential d equals the sum $(pr_1^*)_* \circ d_1 + (pr_2^*)_* \circ d_2$ of the partial differentials; the iterated mixed second derivatives coincide via the isomorphism $L(E, L(F, G)) \cong L(F, L(E, G))$ which is a consequence of 4.2.*

PROOF. Necessity follows by what has been remarked above together with the symmetry of iterated derivatives stated in 4.4. For sufficiency, consider the map $\tilde{d}f \in \mathcal{C}^\infty(E \times F, L(E \times F, G))$ defined by $\tilde{d}f(x)(v_1, v_2) := d_1 f(x)(v_1) + d_2 f(x)(v_2)$. Then obviously for fixed x the map $(t, v) \mapsto \tilde{d}f(x + tv)(v)$ is smooth from $[0, 1] \times E \times F \to G$ and hence can be viewed as an element of $\mathcal{C}^\infty([0, 1], \mathcal{C}^\infty(E \times F, G))$. By [**28**], 2.7, a smooth curve is Riemann integrable, the Riemann integral leads again into $\mathcal{C}^\infty(E \times F, G)$ and commutes with the application of smooth linear maps. It follows that the map

$$v \mapsto f(x) + \int_0^1 \tilde{d}f(x + tv)(v) dt$$

is smooth on $E \times F$ and it suffices to verify that the expression on the right hand side equals $f(x + v)$ in order to obtain smoothness of f on $E \times F$. For this, note that for each fixed segment $[x, x + v]$, we can recover the claimed identity from the finite dimensional one by composing the restriction of f to the segment with bounded linear functionals. □

The *differentiable uniform boundedness principle* (see 4.4.7 in [**22**]) constitutes an extremely useful tool for testing smoothness of linear maps into spaces of smooth functions:

THEOREM 4.7. *Let E, F, G be locally convex spaces, E, G complete. A linear map $E \to \mathcal{C}^\infty(F, G)$ is smooth if and and only if its composition with the evaluation ev_x for each $x \in F$ is smooth.*

If we endow the space $\mathcal{C}^\infty(X, \mathbb{R})$ (in the present work, X will be one of the spaces $\mathcal{D}(\Omega), \mathcal{D}(\Omega) \times \Omega, \mathcal{A}_0(\Omega) \times \Omega$ or $\mathcal{A}_0(\mathbb{R}^s) \times \Omega$) with the topology of uniform convergence on compact subsets in each derivative, i.e., in each iterated differential separately, then by considering the corresponding seminorms one sees that taking the differential constitutes a continuous linear operation. To be precise, the space

$\mathcal{A}_0(\mathbb{R}^s)$ is not a linear space itself but the affine image of the closed linear subspace $E := \mathcal{A}_{00}(\mathbb{R}^s) \subseteq \mathcal{D}(\mathbb{R}^s)$ and may be identified with the latter. A map on $\mathcal{A}_0(\mathbb{R}^s)$ is then said to be smooth if it is the pullback of a smooth map on E under the affine isomorphism. We say that the smooth structure on $\mathcal{A}_0(\mathbb{R}^s)$ is induced by its isomorphism with E. This is a simple example of the notion of a *smooth space* as introduced in [**22**]. Locally convex spaces may be viewed as smooth spaces with a compatible linear structure.

PROPOSITION 4.8. *The following maps (to be defined in chapter 5) are smooth: The linear maps $S_\varepsilon : \mathcal{D}(\mathbb{R}^s) \to \mathcal{D}(\mathbb{R}^s)$, $T_x : \mathcal{D}(\mathbb{R}^s) \to \mathcal{D}(\mathbb{R}^s)$ and $(\varphi, x) \mapsto \bar{\mu}^X(\varphi, x)$, $R \mapsto \hat{\mu}^X R$ ($X \in \{C, J\}$), as well as the non-linear maps S, T, $x \mapsto T_x$, $x \mapsto T_x\varphi$, $(\varphi, x) \mapsto T_x\varphi$.*

PROOF. Smoothness of $S_\varepsilon, T_x, \bar{\mu}^X, \hat{\mu}^X$ follows by our remarks preceding 4.1 and following 4.5, respectively, as each of these maps is essentially a pullback of a smooth map by definition. As the map $S : (\varepsilon, \varphi) \mapsto S_\varepsilon\varphi)$ is linear in φ, it follows by the exponential law 4.2 and the uniform boundedness principle 4.7 that S is smooth iff it is separately smooth, i.e., if and only if the maps S_ε and $(\varepsilon \mapsto S_\varepsilon\varphi)$ are smooth. While smoothness of the former is already established, the latter is a curve which is obviously smooth off 0 and we are done. In a similar fashion, we obtain smoothness of T and all the maps associated with it. □

CHAPTER 5

C- and J-formalism

Colombeau in [**10**] and in [**13**] (together with Meril) on the one hand and Jelínek in [**26**] on the other hand used different, yet equivalent formalisms to describe their respective constructions of Colombeau algebras: For embedding the space $\mathcal{D}'(\mathbb{R}^s)$ of distributions on $\mathbb{R}^s$ into the space $\mathcal{E}_M(\mathbb{R}^s)$ of representatives of generalized functions, they chose different (linear injective) maps which we denote by ι^C ([**10**], [**13**]) and ι^J ([**26**], compare also [**9**]), respectively. On a distribution given by a smooth function f on $\mathbb{R}^s$, ι^C and ι^J are defined by

$$(5.1) \qquad (\iota^C f)(\varphi, x) := \int f(y)\varphi(y-x)\,dy$$

resp.

$$(5.2) \qquad (\iota^J f)(\varphi, x) := \int f(y)\varphi(y)\,dy.$$

Here, φ denotes a test function from the subspace $\mathcal{A}_0(\mathbb{R}^s)$ of $\mathcal{D}(\mathbb{R}^s)$ while $x \in \mathbb{R}^s$. There are good reasons for either of these choices of the embedding; we are going to discuss their respective merits below. In this chapter we show that both formalisms are actually equivalent and establish a translation formalism allowing to change from one setting to the other at any stage of the presentation.

DEFINITION 5.1. For $\varepsilon \in I$ and $x \in \mathbb{R}^s$ define the following operators:

$$(5.3) \qquad T_x : \mathcal{D}(\mathbb{R}^s) \ni \varphi \;\mapsto\; T_x\varphi := \varphi(.\,-x) \in \mathcal{D}(\mathbb{R}^s)$$

$$(5.4) \qquad S_\varepsilon : \mathcal{D}(\mathbb{R}^s) \ni \varphi \;\mapsto\; S_\varepsilon\varphi := \frac{1}{\varepsilon^s}\varphi\!\left(\frac{\cdot}{\varepsilon}\right) \in \mathcal{D}(\mathbb{R}^s)$$

$$(5.5) \qquad S : (0,\infty) \times \mathcal{D}(\mathbb{R}^s) \ni (\varepsilon, \varphi) \;\mapsto\; S_\varepsilon \varphi \in \mathcal{D}(\mathbb{R}^s)$$

$$(5.6) \qquad T : \mathcal{D}(\mathbb{R}^s) \times \mathbb{R}^s \ni (\varphi, x) \;\mapsto\; T(\varphi, x) := (T_x\varphi, x) \in \mathcal{D}(\mathbb{R}^s) \times \mathbb{R}^s$$

$$(5.7) \qquad S^{(\varepsilon)} : \mathcal{D}(\mathbb{R}^s) \times \mathbb{R}^s \ni (\varphi, x) \;\mapsto\; S^{(\varepsilon)}(\varphi, x) := (S_\varepsilon\varphi, x) \in \mathcal{D}(\mathbb{R}^s) \times \mathbb{R}^s.$$

T_x and S_ε are linear. All the operators introduced in the preceding definition are one-one and onto; moreover, they are continuous and smooth with respect to the natural topologies (see chapter 4).

In a next step, we take (5.1) and (5.2) as a starting point for the determination of suitable domains for representatives of generalized functions on an open subset Ω of $\mathbb{R}^s$: Assuming $x \in \Omega$ in (5.1) and (5.2), it is immediate that in (5.2) φ has to have its support in Ω, whereas for (5.1) to be well-defined for any smooth function f on Ω, the support of φ must be contained in $\Omega - x$. This motivates the introduction of the following sets:

5. C- AND J-FORMALISM

DEFINITION 5.2. Let $\varepsilon \in I$.
$$U(\Omega) := T^{-1}(\mathcal{A}_0(\Omega) \times \Omega) \qquad = \{(\varphi, x) \in \mathcal{A}_0(\mathbb{R}^s) \times \Omega \mid \operatorname{supp} \varphi \subseteq \Omega - x\}$$
$$U_\varepsilon(\Omega) := (S^{(\varepsilon)})^{-1}(U(\Omega)) =$$
$$= (TS^{(\varepsilon)})^{-1}(\mathcal{A}_0(\Omega) \times \Omega) = \{(\varphi, x) \in \mathcal{A}_0(\mathbb{R}^s) \times \Omega \mid \operatorname{supp} \varphi \subseteq \varepsilon^{-1}(\Omega - x)\}$$

The notation $U(\Omega)$ is due to Colombeau ([10], 1.2.1). By definition, the maps $T : U(\Omega) \to \mathcal{A}_0(\Omega) \times \Omega$ and $S^{(\varepsilon)} : U_\varepsilon(\Omega) \to U(\Omega)$ are algebraic isomorphisms in the sense that they are bijective and linear in the first argument. The question of topology, however, is somewhat subtle: Let τ_Ω denote the product of the (LF)-topology of $\mathcal{D}(\Omega)$ and the Euclidean topology on Ω; abbreviate $\tau_{\mathbb{R}^s}$ as τ_0. Then on $\mathcal{A}_0(\Omega) \times \Omega$, the topology τ_Ω without doubt is the appropriate one to consider, rather than (the restriction of) τ_0. For $U(\Omega)$, on the other hand, the topology τ_1 induced by τ_0 and the topology $\tau_2 := T^{-1}\tau_\Omega$ both seem to be natural choices. (Note that τ_1 can be obtained equally as $T^{-1}\tau_0$, due to T being a homeomorphism with respect to τ_0.) As the following example (which can easily be generalized to arbitrary non-trivial open subsets of $\mathbb{R}^s$) shows, τ_1 is strictly coarser than τ_2 in general.

EXAMPLE 5.3. Let $\Omega := \{x \in \mathbb{R} \mid x > -1\}$. Choose $\varphi \in \mathcal{D}(\Omega)$ with $\operatorname{supp} \varphi = [0,1]$ and $\int \varphi = 0$. Pick any $\rho \in \mathcal{A}_0(\Omega)$ such that $\operatorname{supp} \rho \subseteq [1,2]$. Letting $\psi_n := \rho + \frac{1}{n}\varphi(. + \frac{n-1}{n}) \in \mathcal{A}_0(\Omega)$, it is easy to check that $T^{-1}(\psi_n, 0) = (\psi_n, 0) \in U(\Omega)$ tends to $(\rho, 0) \in U(\Omega)$ with respect to τ_1, yet is not convergent (in fact, not even bounded) with respect to τ_2.

The situation is similar in the case of $U_\varepsilon(\Omega)$: Apart from the topology $\tau_{1,\varepsilon}$ induced by the topology τ_0 of $\mathcal{A}_0(\mathbb{R}^s) \times \mathbb{R}^s$ via inclusion, the natural topology τ_Ω of $\mathcal{A}_0(\Omega) \times \Omega$ via $TS^{(\varepsilon)}$ induces a topology $\tau_{2,\varepsilon}$ which, in general, is strictly finer than $\tau_{1,\varepsilon}$.

Now, in order to have the respective formalisms of Colombeau and Jelínek equivalent, we want $T : U(\Omega) \to \mathcal{A}_0(\Omega) \times \Omega$ and $S^{(\varepsilon)} : U_\varepsilon(\Omega) \to U(\Omega)$ to be also topological isomorphisms (hence diffeomorphisms). This amounts to endowing $U(\Omega)$ and $U_\varepsilon(\Omega)$ with the topologies τ_2 resp. $\tau_{2,\varepsilon}$ induced via T resp. $TS^{(\varepsilon)}$. Thus we adopt the following convention:

Whenever questions of topology (in particular, boundedness) or smoothness on $U(\Omega)$ or $U_\varepsilon(\Omega)$ are discussed, we consider their topologies to be τ_2 resp. $\tau_{2,\varepsilon}$, i.e., those induced by the natural topology of $\mathcal{A}_0(\Omega) \times \Omega$ via T resp. $TS^{(\varepsilon)}$.

To phrase it differently, $U(\Omega)$ can be viewed as (infinite-dimensional) smooth manifold, modelled over $\mathcal{A}_0(\Omega) \times \Omega$, having an atlas consisting of a single chart T. A similar statement is valid for $U_\varepsilon(\Omega)$ and $TS^{(\varepsilon)}$. The importance as well as the subtlety of distinguishing between τ_1 and τ_2 are highlighted in example 5.9 below.

We are now able to introduce the basic spaces of smooth functions on which the construction of diffeomorphism invariant Colombeau algebras is built.

DEFINITION 5.4.

(5.8) $$\mathcal{E}^J(\Omega) := \mathcal{C}^\infty(\mathcal{A}_0(\Omega) \times \Omega)$$
(5.9) $$\mathcal{E}^C(\Omega) := \mathcal{C}^\infty(U(\Omega))$$

By our above choice of topologies, T^* indeed maps $\mathcal{E}^J(\Omega)$ bijectively onto $\mathcal{E}^C(\Omega)$. The next definition shows how the space of distributions on Ω is to be embedded into $\mathcal{E}^J(\Omega)$ resp. $\mathcal{E}^C(\Omega)$.

DEFINITION 5.5. For $u \in \mathcal{D}'(\Omega)$, define
$$\iota^J : \mathcal{D}'(\Omega) \to \mathcal{E}^J(\Omega) \qquad (\iota^J u)(\varphi, x) := \langle u, \varphi \rangle$$
$$\iota^C : \mathcal{D}'(\Omega) \to \mathcal{E}^C(\Omega) \qquad (\iota^C u)(\varphi, x) := \langle u, \varphi(. - x) \rangle$$

By definition, $\iota^C = T^* \circ \iota^J$.

It remains to introduce the respective extensions of partial differentiation from $\mathcal{D}'(\Omega)$ to $\mathcal{E}^C(\Omega)$ resp. $\mathcal{E}^J(\Omega)$ and the respective actions of a diffeomorphism.

DEFINITION 5.6. For $i = 1, \ldots, s$, define
$$D_i^C : \mathcal{E}^C(\Omega) \to \mathcal{E}^C(\Omega) \qquad D_i^C := \partial_i,$$
$$D_i^J : \mathcal{E}^J(\Omega) \to \mathcal{E}^J(\Omega) \qquad D_i^J := (T^*)^{-1} \circ \partial_i \circ T^*,$$

i.e., for $R \in \mathcal{E}^J(\Omega)$, $(\varphi, x) \in \mathcal{A}_0(\Omega) \times \Omega$ we set
$$(D_i^J R)(\varphi, x) := -((\mathrm{d}_1 R)(\varphi, x))(\partial_i \varphi) + (\partial_i R)(\varphi, x).$$

Of course we have to demonstrate that for given $R \in \mathcal{E}^C(\Omega)$ and $(\varphi, x) \in U(\Omega)$, $(D_i^C R)(\varphi, x)$ in fact exists and that $(\varphi, x) \mapsto (D_i^C R)(\varphi, x)$ is smooth on $U(\Omega)$ with respect to τ_2 (and similar for $R \in \mathcal{E}^J(\Omega)$ and D_i^J). This being a non-trivial task—in particular for the case of the innocent-looking map $D_i^C = \partial_i$ [sic!]—requiring some technical prerequisites, we have to defer it to the following chapter.

Commutativity of the following diagram is immediate:

$$\begin{array}{ccc} \mathcal{D}'(\Omega) & \xrightarrow{\partial_i} & \mathcal{D}'(\Omega) \\ \downarrow \iota^C & & \downarrow \iota^C \\ \mathcal{E}^C(\Omega) & \xrightarrow{D_i^C} & \mathcal{E}^C(\Omega) \\ \uparrow T^* & & \uparrow T^* \\ \mathcal{E}^J(\Omega) & \xrightarrow{D_i^J} & \mathcal{E}^J(\Omega) \end{array}$$

DEFINITION 5.7. Let $\mu : \tilde{\Omega} \to \Omega$ be a diffeomorphism. Define
$$\bar{\mu}^J : \mathcal{A}_0(\tilde{\Omega}) \times \tilde{\Omega} \to \mathcal{A}_0(\Omega) \times \Omega$$
$$\bar{\mu}^C : U(\tilde{\Omega}) \to U(\Omega)$$
$$\bar{\mu}_\varepsilon : U_\varepsilon(\tilde{\Omega}) \to U_\varepsilon(\Omega)$$

by
$$\bar{\mu}^J(\tilde{\varphi}, \tilde{x}) := \left((\tilde{\varphi} \circ \mu^{-1}) \cdot |\det D\mu^{-1}|, \mu\tilde{x} \right),$$
$$\bar{\mu}^C(\tilde{\varphi}, \tilde{x}) := \left(T^{-1} \circ \bar{\mu}^J \circ T \right)(\tilde{\varphi}, \tilde{x})$$
$$= \left(\tilde{\varphi}(\mu^{-1}(. + \mu\tilde{x}) - \tilde{x}) \cdot |\det D\mu^{-1}(. + \mu\tilde{x})|, \mu\tilde{x} \right).$$
$$\bar{\mu}_\varepsilon(\tilde{\varphi}, \tilde{x}) := \left((S^{(\varepsilon)})^{-1} \circ T^{-1} \circ \bar{\mu}^J \circ T \circ S^{(\varepsilon)} \right)(\tilde{\varphi}, \tilde{x})$$
$$= \left(\tilde{\varphi}\left(\frac{\mu^{-1}(\varepsilon. + \mu\tilde{x}) - \tilde{x}}{\varepsilon} \right) \cdot |\det D\mu^{-1}(\varepsilon. + \mu\tilde{x})|, \mu\tilde{x} \right).$$

DEFINITION 5.8. Let $\mu : \tilde{\Omega} \to \Omega$ be a diffeomorphism and $\varepsilon \in I$. Define
$$\hat{\mu}^J : \mathcal{E}^J(\Omega) \to \mathcal{E}^J(\tilde{\Omega})$$
$$\hat{\mu}^C : \mathcal{E}^C(\Omega) \to \mathcal{E}^C(\tilde{\Omega})$$
$$\hat{\mu}_\varepsilon : \mathcal{C}^\infty(U_\varepsilon(\Omega)) \to \mathcal{C}^\infty(U_\varepsilon(\tilde{\Omega}))$$

5. C- AND J-FORMALISM

by $\hat{\mu}^J := (\bar{\mu}^J)^*$, $\hat{\mu}^C := (\bar{\mu}^C)^*$, $\hat{\mu}_\varepsilon := (\bar{\mu}_\varepsilon)^*$, i.e.,

$$
\begin{aligned}
(\hat{\mu}^J R)(\tilde{\varphi}, \tilde{x}) &:= R(\bar{\mu}^J(\tilde{\varphi}, \tilde{x})) & (R \in \mathcal{E}^J(\Omega), & \quad (\tilde{\varphi}, \tilde{x}) \in \mathcal{A}_0(\tilde{\Omega}) \times \tilde{\Omega}), \\
(\hat{\mu}^C R)(\tilde{\varphi}, \tilde{x}) &:= R(\bar{\mu}^C(\tilde{\varphi}, \tilde{x})) & (R \in \mathcal{E}^C(\Omega), & \quad (\tilde{\varphi}, \tilde{x}) \in U(\tilde{\Omega})), \\
(\hat{\mu}_\varepsilon R)(\tilde{\varphi}, \tilde{x}) &:= R(\bar{\mu}_\varepsilon(\tilde{\varphi}, \tilde{x})) & (R \in \mathcal{C}^\infty(U_\varepsilon(\Omega)), & \quad (\tilde{\varphi}, \tilde{x}) \in U_\varepsilon(\tilde{\Omega})).
\end{aligned}
$$

For $X \in \{C, J\}$ we obtain

$$
\begin{array}{ccc}
\mathcal{D}'(\Omega) & \xrightarrow{\mu^*} & \mathcal{D}'(\tilde{\Omega}) \\
\downarrow \iota^X & & \downarrow \iota^X \\
\mathcal{E}^X(\Omega) & \xrightarrow{\hat{\mu}^X} & \mathcal{E}^X(\tilde{\Omega})
\end{array}
$$

where for $u \in \mathcal{D}'(\Omega)$, $\mu^* u$ is defined by $\langle \mu^* u, \varphi \rangle := \langle u, (\varphi \circ \mu^{-1}) \cdot |\det D\mu^{-1}| \rangle$ ($\varphi \in \mathcal{D}(\Omega)$) which extends $f \mapsto \mu^* f = f \circ \mu$ for $f \in \mathcal{C}^\infty(\Omega)$.

$$
\begin{array}{ccc@{\qquad}ccc}
U_\varepsilon(\tilde{\Omega}) & \xrightarrow{\bar{\mu}_\varepsilon} & U_\varepsilon(\Omega) & \mathcal{C}^\infty(U_\varepsilon(\tilde{\Omega})) & \xleftarrow{\hat{\mu}_\varepsilon} & \mathcal{C}^\infty(U_\varepsilon(\Omega)) \\
\downarrow S^{(\varepsilon)} & & \downarrow S^{(\varepsilon)} & \uparrow (S^{(\varepsilon)})^* & & \uparrow (S^{(\varepsilon)})^* \\
U(\tilde{\Omega}) & \xrightarrow{\bar{\mu}^C} & U(\Omega) & \mathcal{E}^C(\tilde{\Omega}) & \xleftarrow{\hat{\mu}^C} & \mathcal{E}^C(\Omega) \\
\downarrow T & & \downarrow T & \uparrow T^* & & \uparrow T^* \\
\mathcal{A}_0(\tilde{\Omega}) \times \tilde{\Omega} & \xrightarrow{\bar{\mu}^J} & \mathcal{A}_0(\Omega) \times \Omega & \mathcal{E}^J(\tilde{\Omega}) & \xleftarrow{\hat{\mu}^J} & \mathcal{E}^J(\Omega)
\end{array}
$$

Definitions 5.7 and 5.8 reflect the fact that in Definition (**D5**) of chapter 3, we chose to regard distributions (and, in the sequel, non-linear generalized functions) as generalizations of *functions*, acting as functionals on test densities (compare, e.g., [**25**]). This approach has to be distinguished from constructing distributions as distributional *densities*, acting on test functions (see, e.g., [**20**]).

In the following table, we compare the C-formalism and the J-formalism regarding simplicity of the respective definitions and, in the last item, the degree of familiarity.

C-formalism	Feature	J-formalism
−	domain of basic space $\mathcal{E}(\Omega)$	+
−	smoothness structure	+
−	embedding of $\mathcal{D}'(\Omega)$	+
+	formula for differentiation	−
+	solving differential equations	−
−	action induced by a diffeomorphism	+
+	formula for testing	−
−	generalization to manifolds	+
+	tradition	−

The distribution of the +'s and −'s in the table should be rather obvious by inspecting the corresponding definitions. Due to the absence of a linear structure on a general smooth manifold, it is clear that the C-formalism does not lend itself to a definition of non-linear generalized functions on manifolds based only on intrinsic terms, whereas the J-formalism in fact does permit such a construction; see [**24**].

We conclude this chapter by presenting an example that emphasizes the importance of carefully distinguishing between the topologies τ_1 and τ_2 on $U(\Omega)$.

EXAMPLE 5.9. We will specify an open subset Ω of $\mathbb{R}^2$, a line segment of the form $\Phi(t) := (\varphi, z) + t(\psi, v)$ $(-1 \leq t \leq 1)$ in $U(\Omega)$ and a distribution u on Ω such that
$$\lim_{t \to 0} \frac{(\iota^C u)(\Phi(t)) - (\iota^C u)(\Phi(0))}{t} = \infty.$$

This seems to suggest that in the point (φ, z), the function $\iota^C u$ on $U(\Omega)$ (which ought to be smooth according to our definitions) has no directional derivative with respect to (ψ, v); or, to phrase it differently, that the composition of the functions $\iota^C u$ and Φ (both of which have the appearance of being smooth) is not even differentiable in (φ, z). We will leave the solution to this puzzle for the end of the example. First we give the details of the construction.

Let $\Omega := \{(x, y) \in \mathbb{R}^2 \mid x > -y^2 - 1\}$, $z := (0, 0)$, $v := (0, 1)$. Choose $\rho_1 \in \mathcal{D}(\mathbb{R})$ such that $\operatorname{supp} \rho_1 \subseteq [0, \frac{3}{2}]$ and $\rho_1(x) = \exp(-\frac{1}{x})$ on I. Let $c := \int \rho_1$ and choose $\rho_2 \in \mathcal{D}(\mathbb{R})$ such that $\operatorname{supp} \rho_2 \subseteq [\frac{3}{2}, 2]$ and $\int \rho_2 = 1$. Then $\rho := \rho_1 - c\rho_2$ has its support contained in $[0, 2]$, coincides with $\exp(-\frac{1}{x})$ on I and satisfies $\int \rho = 0$. Pick $\omega \in \mathcal{D}(\mathbb{R})$ with the properties $\operatorname{supp} \omega = [-2, +2]$, $0 \leq \omega \leq 1$ and $\omega \equiv 1$ on $[-1, +1]$. Now define $\psi \in \mathcal{D}(\mathbb{R}^2)$ by $\psi(x, y) := \omega(y) \cdot \rho(x + 1 - y^2)$.

Finally, in order to obtain Φ as defined above, take any $\varphi \in \mathcal{A}_0(\mathbb{R}^2)$ whose support is located at the right hand side of the line given by $x = 6$. It is easy to check that $\Phi(t) = (\varphi + t\psi, z + tv)$ belongs to $U(\Omega)$ for all $t \in [-1, +1]$.

Now there is still u to be defined. To this end, let
$$f(x) := \begin{cases} \frac{1}{x^2} \exp(\frac{1}{x} + \frac{2}{x^2}) & (0 < x < 1) \\ 0 & (x \geq 1) \end{cases}.$$

For $\sigma \in \mathcal{D}(\Omega)$ define the distribution $u \in \mathcal{D}'(\Omega)$ by
$$\langle u, \sigma \rangle := \int_{-1}^{0} f(x+1)\sigma(x, 0)\, dx = \int_{0}^{1} f(x)\sigma(x-1, 0)\, dx.$$

For $0 < |t| \leq 1$, it follows
$$\frac{1}{t}[(\iota^C u)(\varphi + t\psi, z + tv) - (\iota^C u)(\varphi, z)] = \frac{1}{t} t \langle u, \psi(x, y - t) \rangle = \int_{0}^{1} f(x)\rho(x - t^2)\, dx.$$

We will show that for $0 < |t| \leq \frac{1}{\sqrt{2}}$, the value of the last integral can be estimated from below by $\exp(\frac{1}{t^2} - 1) - \exp(1)$, thus tending to infinity for $t \to 0$. Substituting $x = \frac{1}{u}$, $t^2 = \frac{1}{v}$, we obtain
$$\int_{0}^{1} f(x)\rho(x - t^2)\, dx = \int_{1}^{v} e^{u + 2u^2} e^{-\frac{vu}{v-u}}\, du \geq \int_{1}^{v-1} e^{2u^2} e^{-\frac{u^2}{v-u}}\, du \geq \int_{1}^{v-1} e^{2u^2} e^{-u^2}\, du$$
$$\geq \int_{1}^{v-1} e^u\, du = e^{v-1} - e = e^{\frac{1}{t^2} - 1} - e.$$

The apparent inconsistencies mentioned at the beginning of the example dissolve by taking into account that, in fact, both τ_1 and τ_2 are involved in the argument: The statement that $\Phi : [-1,+1] \to U(\Omega)$ is smooth is true only if it refers to τ_1 (the image of any neighborhood of 0 under Φ is even unbounded with respect to τ_2 since the supports of $T(\Phi(t))$ are not contained in any compact subset of Ω around $t = 0$). The statement that $\iota^C u$ is smooth is true only if $U(\Omega)$ is endowed with the topology τ_2 induced by the natural topology τ_Ω of $\mathcal{A}_0(\Omega) \times \Omega$ via T. τ_2 being strictly finer than τ_1, we cannot infer the differentiability of $(\iota^C u) \circ \Phi$ from the actual smoothness properties of $\iota^C u$ resp. Φ. Another way of capturing the problem is by pointing out that (ψ, v) is not a member of the tangent space to $U(\Omega)$ at (φ, z) (in the sense of the following chapter) since $\operatorname{supp} \psi$ is not contained in $\Omega - z = \Omega$.

CHAPTER 6

Calculus on $U_\varepsilon(\Omega)$

The purpose of this chapter is to develop an appropriate framework for defining and handling differentials of any order of a function $f : U_\varepsilon(\Omega) \to \mathbb{C}$ which is smooth with respect to $\tau_{2,\varepsilon}$ (by definition, f is of the form $f_0 \circ T \circ S^{(\varepsilon)}$ where $f_0 \in \mathcal{C}^\infty(\mathcal{A}_0(\Omega) \times \Omega)$). By choosing $\varepsilon = 1$, this includes the case of smooth functions on $U(\Omega)$, i.e., of elements of the basic space $\mathcal{E}^C(\Omega)$. As a matter of fact, the author of [**26**] has payed only minor attention to these questions. However, it should be clear even from a glimpse at chapters 7 and 10, in particular, that a sound definition and a proper handling of the differentials of $R_\varepsilon := R^J \circ T \circ S^{(\varepsilon)} = R^C \circ S^{(\varepsilon)}$ are crucial for the construction of a diffeomorphism invariant Colombeau algebra.

To start with, we discuss an important property of the sets $U_\varepsilon(\Omega)$ which will be fundamental in the sequel at many places. Loosely speaking, every subset of $\mathcal{A}_0(\mathbb{R}^s) \times \Omega$ which is "not too large" finally gets into $U_\varepsilon(\Omega)$ by scaling and does not feel any difference between $\tau_{1,\varepsilon}$ and $\tau_{2,\varepsilon}$. To this end, we introduce the following notation:

DEFINITION 6.1. For every compact subset K of Ω define

$$\begin{aligned}
\mathcal{A}_{0,K}(\Omega) &:= \{\varphi \in \mathcal{A}_0(\Omega) \mid \operatorname{supp}\varphi \subseteq K\}, \\
\mathcal{A}_{00,K}(\Omega) &:= \{\varphi \in \mathcal{A}_{00}(\Omega) \mid \operatorname{supp}\varphi \subseteq K\}, \\
U_K(\Omega) &:= T^{-1}(\mathcal{A}_{0,K}(\Omega) \times \Omega), \\
U_{\varepsilon,K}(\Omega) &:= (S^{(\varepsilon)})^{-1}(U_K(\Omega)).
\end{aligned}$$

By definition, we have

$$\begin{aligned}
U_K(\Omega) &= \{(\varphi, x) \in \mathcal{A}_0(\mathbb{R}^s) \times \Omega \mid \operatorname{supp}\varphi \subseteq K - x\}, \\
U_{\varepsilon,K}(\Omega) &= (S^{(\varepsilon)})^{-1} T^{-1}(\mathcal{A}_{0,K}(\Omega) \times \Omega) \\
&= \{(\varphi, x) \in \mathcal{A}_0(\mathbb{R}^s) \times \Omega \mid \operatorname{supp}\varphi \subseteq \varepsilon^{-1}(K - x)\}.
\end{aligned}$$

Then it is immediate that for $K \subset\subset \Omega$, the topologies on $\mathcal{A}_{0,K}(\Omega)$ inherited from the natural topologies of $\mathcal{A}_0(\mathbb{R}^s)$ and $\mathcal{A}_0(\Omega)$, respectively, coincide. Consequently, on $U_K(\Omega)$ the topologies τ_1 and τ_2 are equal, as are $\tau_{1,\varepsilon}$ and $\tau_{2,\varepsilon}$ on $U_{\varepsilon,K}(\Omega)$.

We now are in a position to complement Definition 5.6 by establishing that the derivation operators D_i^C and D_i^J are in fact well-defined. From the explicit formulas for D_i^C resp. D_i^J one is certainly tempted to view the former as being the simpler one of them since it does not seem to involve infinite-dimensional calculus. Yet appearances are deceiving in this case: Since we have to view $U(\Omega)$ as a manifold modelled over $\mathcal{A}_0(\Omega) \times \Omega$ the only legitimate way of interpreting $(D_i^C R^C)(\varphi, x) = (\partial_i R^C)(\varphi, x)$ is to push forward the curve $t \mapsto (\varphi, x + te_i)$ via T to $\mathcal{A}_0(\Omega) \times \Omega$ and to study the directional derivative of $R^C \circ T^{-1}$ along $c : t \mapsto T(\varphi, x + te_i) = (\varphi(\,.\, - (x + te_i)), x + te_i)$ at $t = 0$.

To this end, first note that for small absolute values of t, c actually takes values in $\mathcal{A}_0(\Omega) \times \Omega$. Moreover, $t \mapsto c(t)$ is a smooth curve in $\mathcal{A}_0(\Omega) \times \Omega$ with respect to τ_0, for the time being, according to Proposition 4.8. Since c maps some interval $[-\delta, +\delta]$ into $\mathcal{A}_{0,K}(\Omega) \times \Omega$ for a suitable $K \subset\subset \Omega$, the restriction of c to $(-\delta, +\delta)$ is smooth even with respect to τ_Ω. Therefore, the directional derivative of $R^C \circ T^{-1}$ along $c: t \mapsto T(\varphi, x + te_i) = (\varphi(\,.\, - (x + te_i)), x + te_i)$ at $t = 0$ exists. Having established existence, we can calculate its value as being given by

$$\lim_{t \to 0} \frac{1}{t}[R \circ T^{-1}(c(t)) - R \circ T^{-1}(c(0))] = \lim_{t \to 0} \frac{1}{t}[R(\varphi, x + te_i) - R(\varphi, x)].$$

Thus the usual formula works for $R^C \in \mathcal{C}^\infty(U(\Omega))$ and $D_i^C = \partial_i$, although $U(\Omega)$ is not a linear space.

Finally, to see that $\partial_i R^C$ is again smooth, we have to convince ourselves that $(\partial_i R^C) \circ T^{-1} = (T^{-1})^* D_i^C R^C = D_i^J (T^{-1})^* R^C = D_i^J (R^C \circ T^{-1})$ is smooth on $\mathcal{A}_0(\Omega) \times \Omega$. Since, by definition, $R^J := R^C \circ T^{-1}$ is smooth on $\mathcal{A}_0(\Omega) \times \Omega$, so are its differential $\mathrm{d}_1(R^C \circ T^{-1})$ and its partial derivative $\partial_i(R^C \circ T^{-1})$ on their respective domains. By Definition 5.6, $D_i^J(R^C \circ T^{-1}) = (D_i^C R^C) \circ T^{-1}$ is smooth which is equivalent to the smoothness of $D_i^C R^C$ on $U(\Omega)$. The smoothness of $D_i^J R^J$ for given $R^J \in \mathcal{E}^J(\Omega)$, on the other hand, is immediate solely by the last part of the argument given above.

Let us return to studying the sets $U_\varepsilon(\Omega)$. For the purpose of reference, the following observation is formulated as a lemma.

LEMMA 6.2. *Let $K \subset\subset L \subset\subset \mathbb{R}^s$ and let B be a subset of $\mathcal{D}(\mathbb{R}^s)$ such that all $\varphi \in B$ have their supports contained in some bounded set. Then there exists $\eta > 0$ such that $\mathrm{supp}\, S_\varepsilon(\varphi) \subseteq L - x$ for all $\varepsilon \leq \eta$ and $\varphi \in B$, $x \in K$.*

PROOF. Set $h := \mathrm{dist}(K, \partial L)$. Then for each $x \in K$, L contains the closed ball $\overline{B}_h(x)$ of radius h around x. If, on the other hand, the compact ball $D := \overline{B}_r(0)$ contains the supports of all $\varphi \in B$ then putting $\eta := \frac{h}{r}$ will do: We have $\mathrm{supp}\, S_\varepsilon(\varphi) + x \subseteq \varepsilon D + x \subseteq L$ for $\varepsilon \leq \eta$, $\varphi \in B$, $x \in K$. □

PROPOSITION 6.3. *Let $K \subset\subset L \subset\subset \Omega$ and let B be a subset of $\mathcal{A}_0(\mathbb{R}^s)$ such that all $\varphi \in B$ have their support contained in some fixed bounded subset of $\mathbb{R}^s$. Then there exists $\eta > 0$ such that $B \times K \subseteq U_{\varepsilon,L}(\Omega)$ for all $\varepsilon \leq \eta$. In particular, $B \times K \subseteq U_\varepsilon(\Omega)$ and the restrictions of $\tau_{1,\varepsilon}$ and $\tau_{2,\varepsilon}$ to $B \times K$ are equal.*

PROOF. L, K and B satisfying the assumptions of Lemma 6.2, we obtain $\eta > 0$ such that $\mathrm{supp}\, S_\varepsilon(\varphi) \subseteq L - x$, i.e., $(\varphi, x) \in U_{\varepsilon,L}(\Omega)$ for all $\varepsilon \leq \eta$ and $\varphi \in B$, $x \in K$. □

The fact that for small ε the topologies $\tau_{1,\varepsilon}$ and $\tau_{2,\varepsilon}$ agree on sets of the form $B \times K$ as above is crucial to get the smoothness properties right when it comes to testing for moderateness resp. negligibility, as we will see.

With these prerequisites at hand, we now are ready to introduce the tangent space of $U_\varepsilon(\Omega)$ and to define differentials of all orders of a smooth function defined on $U_\varepsilon(\Omega)$. From an abstract point of view, the tangent space of $U_\varepsilon(\Omega)$ with respect to $\tau_{2,\varepsilon}$ at the point $(\varphi, x) \in U_\varepsilon(\Omega)$ is isomorphic to $\mathcal{A}_{00}(\Omega) \times \Omega$; to the tangent vector $(\sigma, v) \in \mathcal{A}_{00}(\Omega) \times \mathbb{R}^s$ at $(\rho, x) \in \mathcal{A}_0(\Omega) \times \Omega$ there corresponds the "tangent vector" $(S_{\frac{1}{\varepsilon}} T_{-x}(\sigma + \mathrm{d}\rho \cdot v), v) \in \mathcal{A}_{00}(\mathbb{R}^s) \times \mathbb{R}^s$ at $(\varphi, x) = (T \circ S^{(\varepsilon)})^{-1}(\rho, x) = (S_{\frac{1}{\varepsilon}} T_{-x} \rho, x) \in U_\varepsilon(\Omega)$ where $\mathrm{d}\rho \cdot v$ denotes the directional derivative of ρ with respect to v. The preceding formula is obtained by taking the derivative at $t = 0$ of the smooth curve

$t \mapsto (T \circ S^{(\varepsilon)})^{-1}(\rho + t\sigma, x + tv)$. In this sense, the tangent space to $U_\varepsilon(\Omega)$ at $(\varphi, x) \in U_\varepsilon(\Omega)$ can be identified with the set of all $(\psi, v) \in \mathcal{A}_{00}(\mathbb{R}^s) \times \mathbb{R}^s$ satisfying $\operatorname{supp} \psi \subseteq \frac{\Omega - x}{\varepsilon}$. Note that in this case the kinematic tangent space coincides with the operational one (the space of derivations defined on the smooth functions): In fact, by [28], 28.7., and [21], this is true for the space $\mathcal{D}(\Omega)$ (more generally, for smooth sections with compact support of vector bundles over a manifold) and hence by [21] for its complemented subspace $\mathcal{A}_0(\Omega)$.

Essentially, Proposition 6.3 also applies to tangent vectors:

PROPOSITION 6.4. *Let $K \subset\subset L \subset\subset \Omega$ and let B, C be subsets of $\mathcal{A}_0(\mathbb{R}^s)$ resp. $\mathcal{A}_{00}(\mathbb{R}^s)$ such that all $\omega \in B \cup C$ have their supports contained in a fixed bounded subset of $\mathbb{R}^s$. Then there exists $\eta > 0$ such that $B \times K \subseteq U_{\varepsilon, L}(\Omega)$ and $C \times \mathbb{R}^s$ is contained in the tangent space to $U_\varepsilon(\Omega)$ at (φ, x) for all $(\varphi, x) \in B \times K$.*

The proof is virtually the same as for Proposition 6.3, with B now replaced by $B \cup C$; it even yields $\operatorname{supp} \psi \subseteq \frac{L-x}{\varepsilon}$ for all tangent vectors (ψ, v) with $\psi \in C$.

Now let $f : U_\varepsilon(\Omega) \to \mathbb{C}$ be a function which is smooth with respect to $\tau_{2,\varepsilon}$. Basically, $\operatorname{d}^n f$ ought to be defined on the n-fold tangent space to $U_\varepsilon(\Omega)$, that is, on

$$\mathcal{T}^n U_\varepsilon(\Omega) := \bigsqcup_{(\varphi, x) \in U_\varepsilon(\Omega)} \{(\varphi, x)\} \times \{(\psi, v) \in \mathcal{A}_{00}(\mathbb{R}^s) \times \mathbb{R}^s \mid \operatorname{supp} \psi \subseteq \frac{\Omega - x}{\varepsilon}\}^n.$$

f being assumed as smooth with respect to $\tau_{2,\varepsilon}$ by definition, we cannot use *a priori* the structure of the surrounding space $\mathcal{A}_0(\mathbb{R}^s) \times \Omega$ to define $\operatorname{d}^n f$. Instead, we will decompose $U_\varepsilon(\Omega)$ (which has to be viewed as a manifold modelled over $\mathcal{A}_0(\Omega) \times \Omega$) into a family of subsets which is characteristic for smoothness of a function with respect to $\tau_{2,\varepsilon}$ in the sense that f is smooth on $U_\varepsilon(\Omega)$ if and only if the restriction of f to any of these subsets is smooth, yet this time—due to equality of $\tau_{1,\varepsilon}$ and $\tau_{2,\varepsilon}$ on each of these subsets—either with respect to $\tau_{1,\varepsilon}$ or $\tau_{2,\varepsilon}$. This allows the calculus of $\mathcal{A}_0(\mathbb{R}^s) \times \Omega$ to be applied to f. In particular, differentials of f of any order can be computed already from the restrictions of f to these subsets; the chain rule holds.

For the following, fix $\varepsilon \in I$. We will simply write S in place of $S^{(\varepsilon)}$.

PROPOSITION 6.5. *For given $x \in \Omega$ and $L \subset\subset \Omega$, set $K := K_{x,L} := \frac{L-x}{\varepsilon} (\subset\subset \frac{\Omega-x}{\varepsilon})$; choose a compact set $M := M_L$ and a positive number $h := h_L$ such that $L \subset\subset M \subset\subset \Omega$ and $0 < h < \operatorname{dist}(L, \partial M)$. Finally, set $U := U_{x,L} := B_h(x)$. Then*

(6.1) $$\mathcal{T}^n U_\varepsilon(\Omega) = \bigcup_{\substack{x \in \Omega \\ L \subset\subset \Omega}} \mathcal{A}_{0,K}(\mathbb{R}^s) \times U \times \left(\mathcal{A}_{00,K}(\mathbb{R}^s) \times \mathbb{R}^s\right)^n$$

$$= \bigcup_{\substack{x \in \Omega \\ L \subset\subset \Omega}} \mathcal{T}^n\left(\mathcal{A}_{0,K}(\mathbb{R}^s) \times U\right).$$

In particular, $U_\varepsilon(\Omega) = \bigcup_{\substack{x \in \Omega \\ L \subset\subset \Omega}} \mathcal{A}_{0,K}(\mathbb{R}^s) \times U$. Moreover, each set $\mathcal{A}_{0,K}(\mathbb{R}^s) \times U$ is contained in the corresponding set $U_{\varepsilon, M}(\Omega)$.

PROOF. For $y \in U$, we have $K = \frac{L + (y-x) - y}{\varepsilon} \subseteq \frac{M-y}{\varepsilon}$. Consequently, $\mathcal{A}_{0,K}(\mathbb{R}^s) \times U$ is a subset of $U_{\varepsilon, M}(\Omega) (\subseteq U_\varepsilon(\Omega))$ and any $(\psi, v) \in \mathcal{A}_{00,K}(\mathbb{R}^s) \times \mathbb{R}^s$ belongs to the tangent space of $U_\varepsilon(\Omega)$ at (φ, y) for arbitrary $(\varphi, y) \in \mathcal{A}_{0,K}(\mathbb{R}^s) \times U$. Conversely, if $(\varphi, x) \in U_\varepsilon(\Omega)$ and vectors (ψ_i, v_i) $(i = 1 \ldots, n)$ tangent to $U_\varepsilon(\Omega)$ at (φ, x) are given, set $L := x + \varepsilon \cdot (\operatorname{supp}\varphi \cup \bigcup_{i=0}^{n} \operatorname{supp}\psi_i)$. Noting that $L \subset\subset \Omega$, we obtain $(\varphi, x) \in \mathcal{A}_{0,K_{x,L}}(\mathbb{R}^s) \times U_{x,L}$ and $(\psi_i, v_i) \in \mathcal{A}_{00,K_{x,L}}(\mathbb{R}^s) \times \mathbb{R}^s$. This establishes (6.1). The last statement of the proposition has already been shown above. □

THEOREM 6.6. *Let $f : U_\varepsilon(\Omega) \to \mathbb{C}$. f is smooth (with respect to $\tau_{2,\varepsilon}$) if and only if the restriction of f to every set $\mathcal{A}_{0,H}(\mathbb{R}^s) \times V$ is smooth with respect to $\tau_{1,\varepsilon}$ where $H \subset\subset \mathbb{R}^s$, V is an open subset of Ω and $\mathcal{A}_{0,H}(\mathbb{R}^s) \times V$ is contained in $U_{\varepsilon,N}(\Omega)$ for some $N \subset\subset \Omega$.*

PROOF. To begin with, note that it follows from $\mathcal{A}_{0,H}(\mathbb{R}^s) \times V \subseteq U_{\varepsilon,N}(\Omega)$ that $\tau_{1,\varepsilon}$ and $\tau_{2,\varepsilon}$ agree on $\mathcal{A}_{0,H}(\mathbb{R}^s) \times V$. Assuming f to be smooth with respect to $\tau_{2,\varepsilon}$ it is now clear that its restriction to any set $\mathcal{A}_{0,H}(\mathbb{R}^s) \times V$ is smooth with respect to $\tau_{1,\varepsilon}$.

Conversely, suppose that f is smooth with respect to $\tau_{1,\varepsilon}$ on any set $\mathcal{A}_{0,K}(\mathbb{R}^s) \times U$ as defined in Proposition 6.5. Let $L_1 \subset\subset \Omega$ and $x \in \Omega$. In a first step, we show that there exists an open neighborhood V_x of x such that $(T \circ S)^{-1}(\mathcal{A}_{0,L_1}(\Omega) \times V_x) \subseteq \mathcal{A}_{0,K}(\mathbb{R}^s) \times U$ for some $K = K_{x,L}$, $U = U_{x,L}$: Choose L as to satisfy $L_1 \subset\subset L \subset\subset \Omega$ and define K, M, h, U as it had been done in Proposition 6.5; finally, set $V_x := B_r(x)$ where $r := \min(h, \operatorname{dist}(L_1, \partial L))$. For $\varphi \in \mathcal{A}_{0,L_1}(\Omega)$ and $y \in V_x$ it now follows that $y \in U$ and $\operatorname{supp} S_{\frac{1}{\varepsilon}} T_{-y} \varphi = \frac{1}{\varepsilon} \cdot (\operatorname{supp}\varphi - y) \subseteq \frac{L_1 - y}{\varepsilon} = \frac{L_1 + (x - y) - x}{\varepsilon} \subseteq \frac{L - x}{\varepsilon} = K$. Altogether, $(T \circ S)^{-1}(\varphi, y) \in \mathcal{A}_{0,K}(\mathbb{R}^s) \times U$. By assumption and due to the inclusion relation just shown, $f \circ (T \circ S)^{-1}$ is smooth on every set $\mathcal{A}_{0,L_1}(\Omega) \times V_x$. Since $L_1 \subset\subset \Omega$ and $x \in \Omega$ have been arbitrary, $f \circ (T \circ S)^{-1}$ is smooth on the whole of $\mathcal{A}_0(\Omega) \times \Omega$ according to (an obvious modification of) Theorem 4.1. By definition, f is smooth. □

An inspection of the preceding proof shows that it is even sufficient for the smoothness of f that the condition stated in the theorem is satisfied for all $H = K_{x,L}$, $V = U_{x,L}$ where $H = K_{x,L}$, $V = U_{x,L}$ are defined as in Proposition 6.5.

Theorem 6.6 allows us to define $d^n f$ on every set $\mathcal{A}_{0,H}(\mathbb{R}^s) \times V$ as above which is contained in some set of the form $U_{\varepsilon,N}(\Omega)$ with $N \subset\subset \Omega$. Since the sets $\mathcal{A}_{0,H}(\mathbb{R}^s) \times V$ cover $U_\varepsilon(\Omega)$, the differentials of f are defined globally on $U_\varepsilon(\Omega)$. Note that, in fact, they are well-defined: Let $(\varphi, x); (\psi_1, v_1), \ldots, (\psi_n, v_n)$ be a set of data with

(6.2) $(\varphi, x) \in U_\varepsilon(\Omega)$, $v_i \in \mathbb{R}^s$, $\psi_i \in \mathcal{A}_{00}(\mathbb{R}^s)$, $\operatorname{supp}\psi_i \subseteq \frac{\Omega - x}{\varepsilon}$ $(i = 1, \ldots, n)$,

which is a member of $\mathcal{T}^n(\mathcal{A}_{0,H_j}(\mathbb{R}^s) \times V_j)$ both for $j = 0$ and 1; then either way of restricting f to $\mathcal{A}_{0,H_j}(\mathbb{R}^s) \times V_j$ gives the same value for $d^n f$ evaluated at these data as the restriction of f to $\mathcal{A}_{0,H}(\mathbb{R}^s) \times V$ would produce, where $H := H_1 \cap H_2$ and $V := V_1 \cap V_2$. In the particular case where f is the restriction of some $\tilde{f} \in \mathcal{C}^\infty(\mathcal{A}_0(\mathbb{R}^s) \times \Omega)$ to $U_\varepsilon(\Omega)$ then, of course, the differentials of f will agree with the restriction of the differentials of $\tilde{f}$ to the corresponding tangent spaces.

Now it is clear that the chain rule holds for any composition of the form $f \circ \Phi$ where $\Phi : W \to U_\varepsilon(\Omega)$ is a smooth function (no matter if with respect to $\tau_{1,\varepsilon}$ or $\tau_{2,\varepsilon}$) such that the domain W of Φ can be covered by a family W_ι of open sets with the property that for each ι, $\Phi(W_\iota)$ is a subset of a suitable set $\mathcal{A}_{0,H}(\mathbb{R}^s) \times V$ being, in turn, contained in some $U_{\varepsilon,N}(\Omega)$. This is exactly the situation we will meet in chapter 7 when constructing the diffeomorphism invariant Colombeau algebra.

REMARK 6.7. In this final remark we drop the assumption of $\varepsilon \in I$ being fixed: We demonstrate that for $R_\varepsilon := R \circ S^{(\varepsilon)}$ with $R \in \mathcal{E}^C(\Omega) = \mathcal{C}^\infty(U(\Omega))$ and for a given family of sets of data $(\varphi, x); (\psi_1, v_1), \ldots, (\psi_n, v_n)$ for which the supports of all $\varphi \in \mathcal{A}_0(\mathbb{R}^s)$ and $\psi_1, \ldots, \psi_n \in \mathcal{A}_{00}(\mathbb{R}^s)$ occurring in this family are contained in a fixed bounded set and x ranges over some compact subset of Ω (v_i being arbitrary from $\mathbb{R}^s$), $\mathrm{d}^n R_\varepsilon(\varphi, x)((\psi_1, v_1), \ldots (\psi_n, v_n))$ is defined for all sufficiently small ε. This will be the typical situation in the applications which are to come along in the sequel. To this end, let a subset B of $\mathcal{D}(\mathbb{R}^s)$ be given such that all $\omega \in B$ have their support contained in some fixed bounded set, say, a closed ball $D := \overline{B}_r(0)$. Choose L satisfying $K \subset\subset L \subset\subset \Omega$ and let $0 < \varepsilon_0 < \frac{1}{r}\mathrm{dist}(L, \partial\Omega)$. For $\varepsilon \leq \varepsilon_0$, set $M := L + \overline{B}_{r\varepsilon}(0)$; we have $M \subset\subset \Omega$. Then $\mathcal{A}_{0,D}(\mathbb{R}^s) \times L^\circ \subseteq U_{\varepsilon,M}(\Omega)$ since $\mathrm{supp}\,\varphi \subseteq D$ and $x \in L$ imply $\mathrm{supp}\,T_x S_\varepsilon \varphi \in \overline{B}_{r\varepsilon}(x) \subseteq M$. Consequently, R_ε is smooth with respect to $\tau_{1,\varepsilon}$ on $\mathcal{A}_{0,D}(\mathbb{R}^s) \times L^\circ$. Moreover, $\mathcal{A}_{00,D}(\mathbb{R}^s)$ is a subset of the tangent space of $U_\varepsilon(\Omega)$ at $(\varphi, x) \in \mathcal{A}_{0,D}(\mathbb{R}^s) \times L$. Hence we conclude that for all $\varepsilon \leq \varepsilon_0$, $\mathrm{d}^n R_\varepsilon(\varphi, x)((\psi_1, v_1), \ldots (\psi_n, v_n))$ is defined for all $x \in K$ (or even $x \in L^\circ$), $\varphi \in B \cap \mathcal{A}_0(\mathbb{R}^s)$, $\psi_1, \ldots, \psi_n \in B \cap \mathcal{A}_{00}(\mathbb{R}^s)$ and $v_1, \ldots, v_n \in \mathbb{R}^s$.

CHAPTER 7

Construction of a diffeomorphism invariant Colombeau algebra

The aim of this chapter is to complete Jelínek's approach to constructing a diffeomorphism invariant Colombeau algebra, guided by the blueprint sketched in chapter 3. Contrary to [26], we base our presentation on the C-formalism—for the convenience of those readers who are acquainted best with the notation used in [10] and [13]. Nevertheless, at each stage it should be possible without difficulty to switch to the J-formalism of [26], due to the translation apparatus described in chapter 5.

Apart from closing a gap in Jelínek's construction we supply those parts of the respective arguments which have been included neither in [13] nor in [26], yet which—according to our view—deserve due attention to be given. This applies, in particular, to (**T4**), (**T5**) and (**T6**). We also include a counterexample showing that, in particular, the question of smoothness of the objects involved in the construction in fact requires a careful treatment.

7.1. The basis for the definition of the algebra

The "basic space" $\mathcal{E}^C(\Omega) = \mathcal{C}^\infty(U(\Omega))$ and the embedding $\iota^C : \mathcal{D}'(\Omega) \to \mathcal{E}^C(\Omega)$ have already been introduced in Definition 5.5. To complete (**D1**), it remains to introduce σ:

DEFINITION 7.1. Let $\sigma : \mathcal{C}^\infty(\Omega) \to \mathcal{E}^C(\Omega)$ be the map defined by

$$(\sigma(f))(\varphi, x) := f(x) \qquad (f \in \mathcal{C}^\infty(\Omega),\ (\varphi, x) \in U(\Omega)).$$

Also, ((**D2**)) has already been taken care of in Definition 5.6. In the next step, we define the subspaces of moderate and negligible members of $\mathcal{E}^C(\Omega)$, respectively. From now on we will make free use of the convention that each (in)equality (E) involving $R(S_\varepsilon \varphi, x)$ with any arguments in place of φ and x is to be understood as "$R(S_\varepsilon \varphi, x)$ is defined [i.e., $(\varphi, x) \in U_\varepsilon(\Omega)$, i.e., $\mathrm{supp}\,\frac{1}{\varepsilon^s}\varphi(\frac{\cdot}{\varepsilon}) \subseteq \Omega - x$] and (E) holds".

DEFINITION 7.2. (**D3**) ([26], 8.) Let $R \in \mathcal{E}^C(\Omega)$. R is called moderate if the following condition is satisfied:

$$\forall K \subset\subset \Omega\ \forall \alpha \in \mathbb{N}_0^s\ \exists N \in \mathbb{N}\ \forall \phi \in \mathcal{C}_b^\infty(I \times \Omega, \mathcal{A}_0(\mathbb{R}^s)) :$$

$$\partial^\alpha (R(S_\varepsilon \phi(\varepsilon, x), x)) = O(\varepsilon^{-N}) \qquad (\varepsilon \to 0)$$

uniformly for $x \in K$. The set of all moderate elements of $R \in \mathcal{E}^C(\Omega)$ will be denoted by $\mathcal{E}_M^C(\Omega)$.

There are several (mutually equivalent) variants of the above condition defining moderateness; six of them are listed below in Theorem 10.5. The formulation in Definition 7.2 is condition (C) in that theorem. Actually, Jelínek has chosen condition (A) of Theorem 10.5 for defining moderateness in [**26**], 8.

DEFINITION 7.3. (**D4**) ([**26**], 18.)
Let $R \in \mathcal{E}_M^C(\Omega)$. R is called negligible if the following condition (which, following [**26**], will be denoted by $(3°)$) is satisfied:

$$\forall K \subset\subset \Omega \ \forall \alpha \in \mathbb{N}_0^s \ \forall n \in \mathbb{N} \ \exists q \in \mathbb{N} \ \forall \phi \in \mathcal{C}_b^\infty(I \times \Omega, \mathcal{A}_q(\mathbb{R}^s)) :$$

$$\partial^\alpha(R(S_\varepsilon\phi(\varepsilon, x), x)) = O(\varepsilon^n) \qquad (\varepsilon \to 0)$$

uniformly for $x \in K$. The set of all negligible elements of $R \in \mathcal{E}^C(\Omega)$ will be denoted by $\mathcal{N}^C(\Omega)$.

Observe that in the preceding definition R is presupposed to be moderate in the sense of Definition 7.2. Also for the condition given in Definition 7.3 (without assuming R to be moderate) there are several equivalent reformulations (see Theorem 10.6).

At this point it might be useful to observe that in the framework of the J-formalism, the tests of (**D3**) and (**D4**) have to be performed with $R(T_x S_\varepsilon \phi(\varepsilon, x), x)$ in place of $R(S_\varepsilon \phi(\varepsilon, x), x)$, due to T^* being the appropriate bijection between $\mathcal{E}^J(\Omega)$ and $\mathcal{E}^C(\Omega)$: If $R^C = T^* R^J$ then $R^C(S_\varepsilon\phi(\varepsilon, x), x) = R^J(T_x S_\varepsilon \phi(\varepsilon, x), x)$; thus R^C is moderate resp. negligible in the C-frame if and only if R^J is moderate resp. negligible in the J-frame.

In order to make (**D3**) and (**D4**) meaningful, $R(S_\varepsilon \phi(\varepsilon, x), x)$ has to depend in a smooth way on x. This statement, though looking innocent at first glance, hides a rather delicate question: The assumptions on R resp. ϕ to be smooth refer to different topologies resp. bornologies. In fact, as Example 7.18 shows, it can happen that $R(S_\varepsilon \phi(\varepsilon, x), x)$ is not even locally bounded as a function of x for fixed, yet arbitrarily small ε. This aspect has been treated neither in [**13**] nor in [**26**]. To put things right, the following argument is needed:

Given $K \subset\subset \Omega$, choose L such that $K \subset\subset L \subset\subset \Omega$. From the boundedness of ϕ we conclude that all $\phi(\delta, x)$ with $\delta \in I$, $x \in L$ have their support contained in a suitable fixed bounded subset of $\mathbb{R}^s$. According to Proposition 6.3 there exists $\eta > 0$ such that for all $\varepsilon \leq \eta$, $W := \{\phi(\delta, x) \mid \delta \in I, x \in L\} \times L$ is a subset of $U_\varepsilon(\Omega)$ and the respective restrictions of $\tau_{1,\varepsilon}$ and $\tau_{2,\varepsilon}$ to W are equal. Consequently, also the restrictions of τ_1 and τ_2 to $S^{(\varepsilon)}(W) = \{S_\varepsilon\phi(\delta, x) \mid \delta \in I, x \in L\} \times L$ agree. Let $L°$ denote the interior of L. On the set $(0, \eta) \times L°$ (which is open in $I \times \Omega$ and contains $(0, \frac{\eta}{2}] \times K$), the map $(\varepsilon, x) \mapsto (S_\varepsilon\phi(\varepsilon, x), x)$ is smooth with respect to τ_1 by assumption, hence also with respect to τ_2. By definition, $R \in \mathcal{C}^\infty(U(\Omega))$ amounts to R being smooth with respect to τ_2. Setting $\varepsilon_0 := \frac{\eta}{2}$, we obtain that $R(S_\varepsilon\phi(\varepsilon, x), x)$ is a smooth function of (ε, x) on the open neighborhood $(0, \eta) \times L°$ of $(0, \varepsilon_0] \times K$ which, finally, makes the test conditions in (**D3**) and (**D4**) actually meaningful.

In this sense, we have to extend the convention we made immediately preceding (**D3**) by requiring that whenever derivatives of a term like $R(S_\varepsilon\phi(\varepsilon, x), x)$ on a set $(0, \varepsilon_0] \times K$ are under consideration, it is to be understood that ε_0 is sufficiently small as to make sure that $(\varepsilon, x) \mapsto R(S_\varepsilon\phi(\varepsilon, x), x)$ is smooth on an open neighborhood of $(0, \varepsilon_0] \times K$.

In chapter 3, definitions (**D3**) and (**D4**) have been viewed as "tests" to be performed on elements R of $\mathcal{E}^C(\Omega)$, investigating their behaviour on so-called "test objects" $\phi(\varepsilon, x)$. In the case at hand, the latter take the form of smooth bounded (in the sense of chapter 2) maps from $I \times \Omega$ into $\mathcal{A}_0(\mathbb{R}^s)$ for testing moderateness of R resp. into $\mathcal{A}_q(\mathbb{R}^s)$ for testing negligibility of R.

Returning to the exclusive use of the C-formalism, we will drop the superscript C in ι^C, D_i^C, $\mathcal{E}^C(\Omega)$, $\mathcal{N}^C(\Omega)$ and $\mathcal{E}_M^C(\Omega)$ from now on. Moreover, note that in the sequel, by $\partial_i = \frac{\partial}{\partial x_i}$ we will always denote the corresponding derivative with respect to x, i.e., for example, $\partial_i \phi(\varepsilon, x) = \frac{\partial}{\partial x_i} \phi(\varepsilon, x)$ which must not be confused with $\frac{\partial}{\partial \xi_i} \phi(\varepsilon, x)(\xi)$.

THEOREM 7.4. **(T1)**

$\quad$ (i) $\iota(\mathcal{D}'(\Omega)) \subseteq \mathcal{E}_M(\Omega)$ $\qquad$ (ii) $\sigma(\mathcal{C}^\infty(\Omega)) \subseteq \mathcal{E}_M(\Omega)$
$\quad$ (iii) $(\iota - \sigma)(\mathcal{C}^\infty(\Omega)) \subseteq \mathcal{N}(\Omega)$ $\qquad$ (iv) $\iota(\mathcal{D}'(\Omega)) \cap \mathcal{N}(\Omega) = \{0\}$.

PROOF. Since this theorem does not occur explicitly in [**26**], we include a proof; we will be more explicit on those aspects which are new, compared to Colombeau algebras already known, and more concise concerning the rest. To start with, let $u \in \mathcal{D}'(\Omega)$, $\phi \in \mathcal{C}_b^\infty(I \times \Omega, \mathcal{A}_0(\mathbb{R}^s))$ and let $K \subset\subset L \subset\subset \Omega$. By the boundedness of ϕ, there exists a bounded subset C of $\mathbb{R}^s$ such that $\operatorname{supp} \phi(\varepsilon, x) \subseteq C$ for all $\varepsilon \in I$, $x \in L$. Consequently, for $x \in K$, $\operatorname{supp} \partial^\alpha \phi(\varepsilon, x) \subseteq C$ for all $\alpha \in \mathbb{N}_0^s$, $\varepsilon \in I$. Since for ε sufficiently small (say, for $\varepsilon \leq \varepsilon_0$), even $K + \varepsilon C$ is contained in L°, we obtain that $\operatorname{supp} \phi(\varepsilon, x)\left(\frac{\cdot - x}{\varepsilon}\right) \subset\subset L^\circ$ for $\varepsilon \leq \varepsilon_0$, $x \in K$. Thus for the values taken by

$$\partial^\alpha((\iota u)(S_\varepsilon \phi(\varepsilon, x), x)) = \left\langle u, \partial^\alpha \frac{1}{\varepsilon^s} \phi(\varepsilon, x)\left(\frac{\cdot - x}{\varepsilon}\right)\right\rangle$$

on $(0, \varepsilon_0] \times K$, only the restriction of u to L° is relevant. Moreover, again by the boundedness of ϕ, each $\partial^\alpha \frac{1}{\varepsilon^s} \phi(\varepsilon, x)\left(\frac{y-x}{\varepsilon}\right)$ is of order at most $\varepsilon^{-|\alpha|-s}$ as $\varepsilon \to 0$, uniformly for $x \in K$, $y \in \mathbb{R}^s$. Finally, integrating the modulus of the latter function over $\mathbb{R}^s$ with respect to y gives values of order at most $\varepsilon^{-|\alpha|}$, uniformly for $x \in K$.

(i) Consider first the case $f \in \mathcal{C}(\Omega)$. Then for $\varepsilon \leq \varepsilon_0$, $|\partial^\alpha((\iota f)(S_\varepsilon \phi(\varepsilon, x), x))|$ is majorized by

$$\sup_L |f| \cdot \int_\Omega \left|\partial^\alpha \frac{1}{\varepsilon^s} \phi(\varepsilon, x)\left(\frac{y-x}{\varepsilon}\right)\right| dy = O(\varepsilon^{-|\alpha|})$$

uniformly for $x \in K$. Since locally every distribution is a derivative of a suitable continuous function, a similar estimate establishes the first inclusion.

(ii) Let $f \in \mathcal{C}^\infty(\Omega)$. Then $\partial^\alpha((\sigma f)(S_\varepsilon \phi(\varepsilon, x), x)) = \partial^\alpha f(x)$ clearly is bounded on any compact set K.

(iii) Consider $(\iota - \sigma)f$ for a given $f \in \mathcal{C}^\infty(\Omega)$. Assume, in addition to the above, that ϕ takes its values in $\mathcal{A}_q(\mathbb{R}^s)$. Then for $\varepsilon \leq \varepsilon_0$ and $x \in K$,

$$\partial^\alpha(\iota f - \sigma f)(S_\varepsilon \phi(\varepsilon, x), x)) =$$

$$\sum_\beta \binom{\alpha}{\beta} \int_{\frac{\Omega-x}{\varepsilon}} \left[(\partial^\beta f)(z\varepsilon + x) - (\partial^\beta f)(x)\right] \partial^{\alpha-\beta}\phi(\varepsilon, x)(z)\, dz.$$

Taylor expansion of each $\partial^\beta f$ up to order q yields that all terms containing a power of ε less or equal to q vanish due to $\partial^{\alpha-\beta}\phi(\varepsilon, x) \in \mathcal{A}_q(\mathbb{R}^s) \cup \mathcal{A}_{q0}(\mathbb{R}^s)$. All the remainder terms are (smooth functions) of order at most ε^{q+1}, uniformly for $x \in K$, $z \in C$. Therefore, $\partial^\alpha(\iota f - \sigma f)(S_\varepsilon \phi(\varepsilon, x), x)) = O(\varepsilon^{q+1})$.

(iv) Suppose $\iota u \in \mathcal{N}(\Omega)$ for some $u \in \mathcal{D}'(\Omega)$. For $K \subset\subset \Omega$ choose $q \in \mathbb{N}$ such that the condition in **(D4)** is satisfied for $\alpha = 0$, $n = 1$. Pick any $\varphi \in \mathcal{A}_q(\mathbb{R}^s)$ and set $\phi(\varepsilon,x) := \varphi$ for all $\varepsilon \in I$, $x \in \Omega$. Then by the negligibility of ιu, $(u * S_\varepsilon \check\varphi)(x) = \langle u, \varepsilon^{-s}\varphi(\varepsilon^{-1}(.-x))\rangle \to 0$ as $\varepsilon \to 0$, uniformly on K. This shows that u, being the weak limit of the smooth regular distributions $(u * S_\varepsilon \check\varphi)$, is equal to 0. □

Also, we immediately get

THEOREM 7.5. **(T2)** ([**26**], 19.) $\mathcal{E}_M(\Omega)$ *is a subalgebra of* $\mathcal{E}(\Omega)$.

THEOREM 7.6. **(T3)** ([**26**], 19.) $\mathcal{N}(\Omega)$ *is an ideal in* $\mathcal{E}_M(\Omega)$.

7.2. The approach taken by J. Jelínek

While the conditions given in Definitions 7.2 and 7.3 are adequate for proving **(T1)**–**(T3)**, we do need appropriate reformulations for establishing the invariance of $\mathcal{E}_M(\Omega)$ and $\mathcal{N}(\Omega)$ under differentiation (**(T4)**, **(T5)**) as well as under the action induced by a diffeomorphism (**(T6)**–**(T8)**). Suitable equivalent conditions allowing to prove **(T4)** and **(T5)**, on the one hand, are given in Theorem 17 and in part $(3°) \Leftrightarrow (2°)$ of Theorem 18 in [**26**], respectively. We will quote these theorems below as 7.12 and 7.13.

To establish diffeomorphism invariance, on the other hand, two problems have to be coped with: First, transformed test objects in general are not defined on the whole of $I \times \Omega$; secondly, the property $\phi(\varepsilon,x) \in \mathcal{A}_q(\mathbb{R}^s)$ (as occurring in Definition 7.3) is not preserved under the action of a diffeomorphism. The first of these aspects, though presenting considerable intricacies, is covered only by a few remarks in [**26**] which, in our view, do not provide a treatment as rigorous as these questions require. The appropriate reformulations of Definitions 7.2 and 7.3 dealing with the poor domains of transformed test objects are provided by (C)$\Leftrightarrow$(Z) of Theorem 10.5 and (C″)$\Leftrightarrow$(Z″) of Corollary 10.7, respectively. In order to cope with the problem of $\phi(\varepsilon,x) \in \mathcal{A}_q(\mathbb{R}^s)$ not being preserved by a diffeomorphism, Jelínek claims in part $(3°) \Leftrightarrow (4°)$ of [**26**], Theorem 18 that $R \in \mathcal{E}_M(\Omega)$ is negligible (condition $(3°)$) if and only if it passes the test on test objects ϕ having only asymptotically vanishing moments of order q on K (condition $(4°)$), as compared to $\phi(\varepsilon,x) \in \mathcal{A}_q(\mathbb{R}^s)$ required by condition $(3°)$. While $(4°) \Rightarrow (3°)$ is obvious, the converse statement is not true (see Example 7.7 below). The error in the proof of $(3°) \Rightarrow (4°)$ consists in passing from terms of the form $\mathrm{d}_1 \partial^\alpha [R_\varepsilon(\dots)]$ to $[\mathrm{d}_1 \partial^\alpha R_\varepsilon](\dots)$ without applying the chain rule with respect to the composition of R_ε with some "inner" function represented by the dots (compare the proof of Theorem 7.9 given in chapter 10).

As a consequence, the construction of a diffeomorphism invariant Colombeau algebra aimed at in [**26**] is not complete in the following sense: Eliminating condition $(4°)$ from Theorem 18 deprives one of the possibility of proving diffeomorphism invariance for the algebra at hand. If, on the other hand, $(4°)$ is accepted as defining membership in $\mathcal{N}(\Omega)$ (provided $R \in \mathcal{E}_M(\Omega)$) then the embedding of $\mathcal{D}'(\Omega)$ into $\mathcal{G}(\Omega)$ does not preserve the product of smooth functions (being considered as regular distributions) even in the most simple cases, as can be seen from part two of Example 7.7 below. To overcome this difficulty, we will present a substitute for condition $(4°)$ (see Theorem 7.9 below) which in fact is equivalent to $(3°)$ under the assumption of moderateness and, moreover, allows to deduce diffeomorphism invariance of $\mathcal{N}(\Omega)$.

EXAMPLE 7.7. (1) Let $\Omega := \mathbb{R}$ and denote by u the regular distribution on $\mathbb{R}$ defined by $\langle u, \varphi \rangle := \int \xi \varphi(\xi)\, d\xi$ ($\varphi \in \mathcal{D}(\mathbb{R})$). According to part (iii) of Theorem 7.4, $R := \iota u - \sigma u$ is a member of $\mathcal{N}(\mathbb{R})$, that is, R is moderate and satisfies the condition specified in Definition 7.3 which is condition (3°) of [**26**], Theorem 18. We are going to show that R in fact violates condition (4°), thereby disproving (3°)$\Rightarrow$(4°). It is immediate from the definitions that R is given by $R(\varphi, x) := \int \xi \varphi(\xi) \, d\xi$ ($\varphi \in \mathcal{A}_0(\mathbb{R})$, $x \in \mathbb{R}$). Set K:=$\{0\}$, $\alpha := 1$ and $n := 2$. For any given $q \in \mathbb{N}$, define a test object ϕ_q by $\phi_q(\varepsilon, x) := \varphi_q + x \cdot \psi_q$ ($0 < \varepsilon \leq 1$, $x \in \mathbb{R}$) where φ_q is an arbitrary fixed member of $\mathcal{A}_q(\mathbb{R})$ and $\psi_q \in \mathcal{A}_{00}(\mathbb{R})$ is chosen as to satisfy $\int \xi \psi_q(\xi) \, d\xi = 1$. Then ϕ_q belongs to $\mathcal{C}_b^\infty(I \times \mathbb{R}, \mathcal{A}_0(\mathbb{R}))$ and, being equal to φ_q on $K = \{0\}$, has asymptotically vanishing moments of order q on K, as required by condition (4°). Yet

$$\partial(R(S_\varepsilon \phi_q(\varepsilon, x), x)) = \partial(\varepsilon \cdot x) = \varepsilon \neq O(\varepsilon^2),$$

no matter how large q is chosen. This manifestly contradicts condition (4°) for the choices of K, α, n made above. We also see that adopting (4°) (together with moderateness, of course) as defining property for $\mathcal{N}(\mathbb{R})$ would invalidate part (iii) of Theorem (**T1**) which is the basis for ι to preserve the product of smooth functions. This is made explicit by the following item.

(2) Define ϕ_q as in Example (1), yet this time requiring both $\int \xi \psi_q(\xi) \, d\xi = 1$ and $\int \xi^2 \psi_q(\xi) \, d\xi = 0$ for ψ_q to be chosen from $\mathcal{A}_{00}(\mathbb{R})$ and, in addition, $\varphi_q \in \mathcal{A}_{\max(2,q)}(\mathbb{R})$. Denoting by f the identity function on $\mathbb{R}$, f can be identified with the distribution u introduced previously. A straightforward calculation yields

$$\bigl(\iota(f) \cdot \iota(f) - \iota(f \cdot f)\bigr)(\varphi, x) = \left(\int \xi \varphi(\xi) \, d\xi \right)^2 - \int \xi^2 \varphi(\xi) \, d\xi$$

where $\varphi \in \mathcal{A}_0(\mathbb{R})$, $x \in \mathbb{R}$. Substituting $S_\varepsilon \phi_q(\varepsilon, x)$ for φ and taking second derivatives at $x = 0$, we obtain

$$\partial^2 \bigl((\iota(f) \cdot \iota(f) - \iota(f \cdot f))(S_\varepsilon \phi_q(\varepsilon, x), x)\bigr)\bigr|_{x=0} = \partial^2 \bigl(\varepsilon^2 \cdot x^2\bigr)\bigr|_{x=0} = 2\varepsilon^2 \neq O(\varepsilon^3)$$

for any $q \in \mathbb{N}$. Since again ϕ_q is a test object having asymptotically vanishing moments of order q on $K = \{0\}$, $\bigl(\iota(f) \cdot \iota(f) - \iota(f \cdot f)\bigr)$ does not satisfy (4°), this time with respect to $K = \{0\}$, $\alpha := 2$, $n := 3$. Consequently, adopting (4°) in place of (3°) as the defining property for negligibility would prevent the restriction of ι to $\mathcal{C}^\infty(\mathbb{R})$ from being an algebra homomorphism.

To complete the prerequisites for establishing (**T4**)–(**T8**) it remains to state the theorem replacing part (3°) $\Leftrightarrow$ (4°) of Theorem 18 of [**26**]. To this end, we introduce the following terminology (which, actually, is taken from Part 2 of this monograph):

DEFINITION 7.8. Let $\phi \in \mathcal{C}_b^\infty(I \times \mathbb{R}, \mathcal{A}_0(\mathbb{R}))$, $K \subset\subset \Omega$, $q \in \mathbb{N}$. ϕ is said to be of type $[\mathrm{A}_l^\infty]_{K,q}$ if all derivatives $\partial_x^\beta \phi(\varepsilon, x)$ ($\beta \in \mathbb{N}_0^s$) have asymptotically vanishing moments of order q on K.

In the preceding definition, "A", "l" and "∞" stand for "asymptotically vanishing moments", "locally" (i.e., only on the particular compact set K under consideration) and "derivatives of all orders".

THEOREM 7.9. *Let $R \in \mathcal{E}_M(\Omega)$. R is negligible, i.e., R satisfies the condition specified in Definition 7.3 if and only if it satisfies the following property (which will be referred to as (4^∞)):*

$\forall K \subset\subset \Omega \, \forall \alpha \in \mathbb{N}_0^s \, \forall n \in \mathbb{N} \, \exists q \in \mathbb{N} \, \forall \phi \in \mathcal{C}_b^\infty(I \times \Omega, \mathcal{A}_0(\mathbb{R}^s))$, ϕ of type $[\mathrm{A}_l^\infty]_{K,q}$:

$$\partial^\alpha(R(S_\varepsilon \phi(\varepsilon, x), x)) = O(\varepsilon^n) \qquad (\varepsilon \to 0)$$

uniformly for $x \in K$.

The proof of the preceding theorem is deferred to chapter 10. Restricting β in the definition of type $[A_1^\infty]_{K,q}$ to the value $0 \in \mathbb{N}_0^s$ turns condition (4^∞) into condition $(4°)$ of Theorem 18 of [**26**].

7.3. Stability under differentiation

Having set up the prerequisites for the remaining part of the construction in the previous section we proceed to establish Theorems (**T4**) and (**T5**).

THEOREM 7.10. (**T4**) *For $R \in \mathcal{E}_M(\Omega)$, $\partial_i R \in \mathcal{E}_M(\Omega)$ $(i = 1, \ldots, s)$.*

THEOREM 7.11. (**T5**) *For $R \in \mathcal{N}(\Omega)$, $\partial_i R \in \mathcal{N}(\Omega)$ $(i = 1, \ldots, s)$.*

Curiously enough, the preceding Theorems are not even mentioned in Jelínek's paper [**26**]. We regard them as highly non-trivial, however: At first glance they might seem obvious since the respective tests ask for a certain behaviour of *all* derivatives $\partial^\alpha(R(\phi(\varepsilon, x), x))$; thus, as one might be tempted to argue, the moderateness or negligibility of R implies the respective property also for each $D_i^C R = \partial_i R$. This reasoning, however, does not take into account the fact that the expression $\partial^\alpha(R(\phi(\varepsilon, x), x))$ used for testing R involves a certain sum of partial derivatives of R multiplied by partial derivatives of $\phi(\varepsilon, x)$, according to the chain rule. There is no easy relation between the respective expressions for ∂^α and $\partial_i \partial^\alpha$ which could be used to draw from the asymptotic behaviour of the former to infer the corresponding property for the latter.

The key tools for establishing Theorems (**T4**) and (**T5**) are Jelínek's Theorem 17 and part $(2°) \Leftrightarrow (3°)$ of Theorem 18 in [**26**]. For their ingenious proofs we refer to the original [**26**]. We presume that the author was completely aware of the rôle Theorems 17 and 18 had to play in that respect, yet for some reasons he decided not to address this issue.

THEOREM 7.12. ([**26**], Theorem 17) *Let $R \in \mathcal{E}(\Omega)$. R is moderate (i.e., a member of $\mathcal{E}_M(\Omega)$) if and only if the following condition is satisfied:*

$$\forall K \subset\subset \Omega \ \forall \alpha \in \mathbb{N}_0^s \ \forall k \in \mathbb{N}_0 \ \exists N \in \mathbb{N} \ \forall B \ (bounded) \subseteq \mathcal{D}(\mathbb{R}^s):$$

$$\partial^\alpha \mathrm{d}_1^k (R \circ S^{(\varepsilon)})(\varphi, x)(\psi_1, \ldots, \psi_k) = O(\varepsilon^{-N}) \qquad (\varepsilon \to 0)$$

uniformly for $x \in K$, $\varphi \in B \cap \mathcal{A}_0(\mathbb{R}^s)$, $\psi_1, \ldots, \psi_k \in B \cap \mathcal{A}_{00}(\mathbb{R}^s)$.

THEOREM 7.13. ([**26**], Theorem 18, $(3°) \Leftrightarrow (2°)$)
Let $R \in \mathcal{E}_M(\Omega)$. R is negligible (i.e., a member of $\mathcal{N}(\Omega)$) if and only if the following condition is satisfied:

$$\forall K \subset\subset \Omega \ \forall \alpha \in \mathbb{N}_0^s \ \forall k \in \mathbb{N}_0 \ \forall n \in \mathbb{N} \ \exists q \in \mathbb{N} \ \forall B \ (bounded) \subseteq \mathcal{D}(\mathbb{R}^s):$$

$$\partial^\alpha \mathrm{d}_1^k (R \circ S^{(\varepsilon)})(\varphi, x)(\psi_1, \ldots, \psi_k) = O(\varepsilon^n) \qquad (\varepsilon \to 0)$$

uniformly for $x \in K$, $\varphi \in B \cap \mathcal{A}_q(\mathbb{R}^s)$, $\psi_1, \ldots, \psi_k \in B \cap \mathcal{A}_{q0}(\mathbb{R}^s)$.

Actually, the respective conditions in Definition 7.3 and Theorem 7.13 (that is, conditions $(3°)$ and $(2°)$ of Theorem 18 of [**26**]) are equivalent even without assuming R to be moderate: The proof is similar to the proof of Theorem 17 of [**26**], taking into account the equivalence of conditions (A′) and (C′) of Theorem 10.6.

It is a remarkable fact that in the condition of Theorem 7.13, k, d_1^k and $\psi_1, \ldots, \psi_k$ can be omitted completely (i.e., only the case $k = 0$ has to be taken into account) without changing its content, provided R is assumed to be moderate

([**26**], Theorem 18, $(1°) \Leftrightarrow (2°)$). We can interpret this heuristically as the fact that the moderateness condition takes care of the derivatives of R to be small in the limit, provided only R itself is small in the appropriate sense. A still stronger result will be shown in Part 2 of this monograph: Provided that R is moderate, it is even possible to omit the x-derivatives in the condition of Theorem 7.13, yielding that $R \in \mathcal{E}_M(\Omega)$ is negligible if and only if the following condition is satisfied:

$$\forall K \subset\subset \Omega \, \forall n \in \mathbb{N} \, \exists q \in \mathbb{N} \, \forall B \text{ (bounded)} \subseteq \mathcal{D}(\mathbb{R}^s) : R(S_\varepsilon \varphi, x) = O(\varepsilon^n) \quad (\varepsilon \to 0)$$

uniformly for $x \in K$, $\varphi \in B \cap \mathcal{A}_q(\mathbb{R}^s)$.

Proof of (T4) and (T5). In order to derive, for example, **(T4)** from Theorem 7.12, assume $R \in \mathcal{E}(\Omega)$ to be moderate, hence to satisfy the condition in 7.12. Let $i \in \{1, \ldots, s\}$; due to $\partial_i(R \circ S^{(\varepsilon)}) = (\partial_i R) \circ S^{(\varepsilon)}$ we obtain $\partial^\alpha d_1^k((\partial_i R) \circ S^{(\varepsilon)}) = \partial^{\alpha + e_i} d_1^k(R \circ S^{(\varepsilon)})$ where e_i denotes the i-th standard unit vector in $\mathbb{R}^s$. From the preceding equation it is immediate that together with the differentials of R, also the differentials of $\partial_i R$ are of order at most ε^{-N} for $\varepsilon \to 0$ in the appropriate sense. Applying Theorem 7.12 once more, we infer the moderateness of $\partial_i R$. The proof of Theorem **(T5)** proceeds along the same lines, this time using Theorem 7.13. □

7.4. Diffeomorphism invariance

For any diffeomorphism $\mu : \tilde{\Omega} \to \Omega$ the requirements of **(D5)** are satisfied for $\bar{\mu}^C : U(\tilde{\Omega}) \to U(\Omega)$ and $\hat{\mu}^C : \mathcal{E}^C(\Omega) \to \mathcal{E}^C(\tilde{\Omega})$ as in Definitions 5.7 and 5.8. Again, we shall omit the superscript C from $\bar{\mu}^C$, $\hat{\mu}^C$.

Our next task is to establish the invariance of test objects under the appropriate action induced by μ. This, of course, is at the very heart of the diffeomorphism invariance of the Colombeau algebra to be constructed. In the end, we must be able to infer the moderateness of $\hat{\mu} R$ from the moderateness of R (and, similarly, for negligibility). Unfortunately, it need not be true in a strict sense that the class of test objects $\phi(\varepsilon, x)$ as in Definitions **(D3)** resp. **(D4)** is invariant under the action of a diffeomorphism: The transformed test objects turn out to be defined not on the whole of $I \times \Omega$ necessarily, i.e., they form a larger class than the original test objects do. Due to $\hat{\mu} R = R \circ \bar{\mu}$, we must start with the (formally stronger) assumption (let us denote it by (Z)) that R is moderate even with respect to that larger class of test objects to reach the conclusion that $\hat{\mu} R$ is moderate in the sense of **(D3)** (condition (C)). However, in Theorem 10.5 we will show that (C) and (Z) are, in fact, equivalent so that, in the end, the property of moderateness is shown to be preserved under the action of a diffeomorphism.

The following heuristic calculation clearly shows which path is to be pursued: Let $\bar{\mu}_\varepsilon$ be defined as in chapter 5. For $\tilde{\phi} \in \mathcal{C}_b^\infty(I \times \tilde{\Omega}, \mathcal{A}_0(\mathbb{R}^s))$ given, we have to determine a function ϕ defined on a suitable subset of $I \times \Omega$ taking values in $\mathcal{A}_0(\mathbb{R}^s)$ as to satisfy the following relation:

$$(\hat{\mu} R)(S_\varepsilon \tilde{\phi}(\varepsilon, \tilde{x}), \tilde{x}) = R(\bar{\mu}(S_\varepsilon \tilde{\phi}(\varepsilon, \tilde{x}), \tilde{x})) = R(\bar{\mu} S^{(\varepsilon)}(\tilde{\phi}(\varepsilon, \tilde{x}), \tilde{x}))$$
$$= R(S^{(\varepsilon)}(S^{(\varepsilon)})^{-1} \bar{\mu} S^{(\varepsilon)}(\tilde{\phi}(\varepsilon, \tilde{x}), \tilde{x})) = R(S^{(\varepsilon)} \bar{\mu}_\varepsilon(\tilde{\phi}(\varepsilon, \tilde{x}), \tilde{x})) = R(S_\varepsilon \phi(\varepsilon, \mu \tilde{x}), \mu \tilde{x})$$

where ϕ is defined implicitly by the requirement of the last equality to hold. Obviously, this is the case if and only if $(\phi(\varepsilon, x), x) = \bar{\mu}_\varepsilon(\tilde{\phi}(\varepsilon, \mu^{-1} x), \mu^{-1} x)$ which, according to 5.7, amounts to (7.1) in Theorem 7.14 below. To carry out the program outlined above, three aspects of ϕ have to be handled simultaneously: domain of definition, smoothness and boundedness.

Starting with the first of these, observe that the right hand side of (7.1) is only defined if ξ is an element of $\frac{\Omega-x}{\varepsilon}$ whereas we would want $\xi \mapsto \phi(\varepsilon,x)(\xi)$ to be a test function on the whole of $\mathbb{R}^s$. For the convenience of the reader, we include what essentially is the argument in [26], Remark 25: The right hand side of (7.1) (viewed as a smooth function on $\frac{\Omega-x}{\varepsilon}$) can be extended to a smooth function on the whole of $\mathbb{R}^s$ by setting it equal to 0 for $\xi \notin \frac{\Omega-x}{\varepsilon}$, provided its support is a compact subset of $\frac{\Omega-x}{\varepsilon}$. This, in turn, is equivalent to $\tilde\phi(\varepsilon,\mu^{-1}x)$ having compact support contained in $\frac{\tilde\Omega-\mu^{-1}x}{\varepsilon}$: Indeed, $\xi \mapsto \frac{\mu^{-1}(\varepsilon\xi+x)-\mu^{-1}x}{\varepsilon}$ maps $\frac{\Omega-x}{\varepsilon}$ diffeomorphically onto $\frac{\tilde\Omega-\mu^{-1}x}{\varepsilon}$. Therefore, the largest natural domain of definition for ϕ is the set D of all $(\varepsilon,x) \in I \times \Omega$ for which $\operatorname{supp}\tilde\phi(\varepsilon,\mu^{-1}x)$ is contained in $\frac{\tilde\Omega-\mu^{-1}x}{\varepsilon}$, i.e., for which $(\tilde\phi(\varepsilon,\mu^{-1}x),\mu^{-1}x) \in U_\varepsilon(\tilde\Omega)$. We do not know the form of D explicitly; however, due to the boundedness of the map $\tilde\phi$, for each given $K \subset\subset \Omega$ there exists $\varepsilon_0 > 0$ (chosen appropriately with respect to the compact set $\mu^{-1}K$ by Proposition 6.3) such that $(\tilde\phi(\varepsilon,\mu^{-1}x),\mu^{-1}x) \in U_\varepsilon(\tilde\Omega)$ for all $(\varepsilon,x) \in (0,\varepsilon_0] \times K$ which amounts to $(0,\varepsilon_0] \times K$ being a subset of D. Summarizing, ϕ is defined at least on "rectangles" of the form $(0,\varepsilon_0] \times K$ as a map taking values in $\mathcal{A}_0(\mathbb{R}^s)$. This settles the problem of the domain of ϕ in a satisfactory way, as we will see shortly.

In the light of Example 5.9 as well as of Example 7.18 at the end of this chapter, it seems advisable to give a careful treatment also to the question of smoothness of ϕ. We defer this to the formal proof of **(T6)**.

THEOREM 7.14. **(T6)** ([26], 25.) *Let $\mu : \tilde\Omega \to \Omega$ be a diffeomorphism. Let $\tilde\phi \in \mathcal{C}_b^\infty(I \times \tilde\Omega, \mathcal{A}_0(\mathbb{R}^s))$ and define $D(\subseteq I \times \Omega)$ by*

$$D := \{(\varepsilon,x) \in I \times \Omega \mid (\tilde\phi(\varepsilon,\mu^{-1}x),\mu^{-1}x) \in U_\varepsilon(\tilde\Omega)\}.$$

For $(\varepsilon,x) \in D$, set $\phi(\varepsilon,x) := \operatorname{pr}_1 \bar\mu_\varepsilon(\tilde\phi(\varepsilon,\mu^{-1}x),\mu^{-1}x))$, i.e.,

$$(7.1) \quad \phi(\varepsilon,x)(\xi) := \tilde\phi(\varepsilon,\mu^{-1}x)\left(\frac{\mu^{-1}(\varepsilon\xi+x)-\mu^{-1}x}{\varepsilon}\right) \cdot |\det D\mu^{-1}(\varepsilon\xi+x)|.$$

Then ϕ satisfies the requirements 1) and 2) specified for test objects in condition (Z) of Theorem 10.5.

If, in addition, all derivatives $\partial_{\tilde x}^\alpha \tilde\phi(\varepsilon,\tilde x)$ have asymptotically vanishing moments of order q on some compact subset $\tilde L$ of $\tilde\Omega$ ($\alpha \in \mathbb{N}_0^s$) then all derivatives $\partial_x^\alpha \phi(\varepsilon,x)$ of the the function ϕ defined by (7.1) have asymptotically vanishing moments of order $\left[\frac{q+1}{2}\right]$ on the (compact) set $L = \mu(\tilde L)$.

PROOF. That ϕ is well-defined on D has already been shown. To establish the smoothness of ϕ on suitable open subsets of $I \times \Omega$, expand $\bar\mu_\varepsilon$ to obtain

$$(\phi(\varepsilon,x),x) = (S^{(\varepsilon)})^{-1} T^{-1} \bar\mu^J T S^{(\varepsilon)}(\tilde\phi(\varepsilon,\mu^{-1}x),\mu^{-1}x)).$$

In a first step, we discuss the smoothness of $\Phi(\varepsilon,x) := TS^{(\varepsilon)}(\tilde\phi(\varepsilon,\mu^{-1}x),\mu^{-1}x))$. Φ involves the maps μ^{-1}, $\tilde\phi$, S and T, all of which are smooth by the results of Proposition 4.8, provided $\mathcal{A}_0(\mathbb{R}^s)$ is endowed with the natural locally convex topology inherited from $\mathcal{D}(\mathbb{R}^s)$. Let $K \subset\subset \Omega$; we are going to show that for suitable $\varepsilon_0 > 0$ and $\tilde M \subset\subset \tilde\Omega$, Φ actually maps some open neighborhood of $(0,\varepsilon_0] \times K$ into $\mathcal{A}_{0,\tilde M}(\tilde\Omega) \times \tilde\Omega$. To this end, choose L, M such that $K \subset\subset L \subset\subset M \subset\subset \Omega$. Set $\tilde L := \mu^{-1}L$, $\tilde M := \mu^{-1}M$ and $h := \operatorname{dist}(L,\partial M)$, $\tilde h := \operatorname{dist}(\tilde L,\partial \tilde M)$.

Due to the boundedness of $\tilde\phi$, there is $r \geq \tilde h$ (> 0) such that the supports of all $\tilde\phi(\varepsilon,\tilde x)$ with $\varepsilon \in I$, $\tilde x \in \tilde L$ are contained in the closed ball $\overline B_r(0)$. Setting $\eta := \frac{\tilde h}{r}$, Proposition 6.3 and a glance at the proof of Lemma 6.2 show that $\tilde\phi(I \times \tilde L) \times \tilde L \subseteq U_{\varepsilon,\tilde M}(\tilde\Omega)$ for all $\varepsilon \leq \eta$. In particular, for all $x \in L$ and $\varepsilon \leq \eta$,

$$(7.2) \qquad (\tilde\phi(\varepsilon,\mu^{-1}x),\mu^{-1}x) \in U_{\varepsilon,\tilde M}(\Omega).$$

Therefore, Φ maps the open set $U := (0,\eta) \times L^\circ$ into $\mathcal{A}_{0,\tilde M}(\tilde\Omega) \times \tilde\Omega$. On the latter, however, the topologies τ_0 and $\tau_{\tilde\Omega}$ introduced in chapter 5 coincide. From the smoothness of the restriction of Φ to U with respect to τ_0 (which was established above) we conclude the smoothness with respect to $\tau_{\tilde\Omega}$. Now we are ready to go on with the proof of the smoothness of $\phi = \mathrm{pr}_1 \circ (S^{(\varepsilon)})^{-1} \circ T^{-1} \circ \bar\mu^J \circ \Phi$, observing that $\bar\mu^J$ is smooth if the domain $\mathcal{A}_0(\tilde\Omega) \times \tilde\Omega$ and the range space $\mathcal{A}_0(\Omega) \times \Omega$ carry the topologies $\tau_{\tilde\Omega}$ and τ_Ω, respectively. (Note that in general, $\bar\mu^J$ is *not* τ_0-τ_0-smooth as can be seen from Example 7.19 below.) Weakening this conclusion by replacing τ_Ω by τ_0 on $\mathcal{A}_0(\Omega) \times \Omega$ and using once more the smoothness of T and S with respect to the usual topology of $\mathcal{A}_0(\mathbb{R}^s)$, we finally obtain that for $\varepsilon_0 := \frac{1}{2}\eta$ (if $\eta = 1$ we may choose $U := I \times L^\circ$, being open in $I \times \Omega$, and $\varepsilon_0 := 1$), ϕ is smooth on the open neighborhood U of $(0,\varepsilon_0] \times K$, as claimed by condition 1) of Theorem 10.5 (Z).

For the proof of boundedness of ϕ, we extend the argument of [**26**], 25., Proposition. Note that, by (7.2) above, ϕ is defined at least on $(0,\eta] \times L$. Let $l := \max(1,\sup\{\|D\mu_{\tilde x}\| \mid \tilde x \in \tilde M\})$. Then for $\tilde x \in \tilde L$, we have

$$(7.3) \qquad \|\tilde x - \tilde y\| \leq \tilde h \Rightarrow \|\mu\tilde x - \mu\tilde y\| \leq l\|\tilde x - \tilde y\|,$$

due to $\overline B_{\tilde h}(\tilde x) \subseteq \tilde M$. $\overline B_r(0)$ containing the support of every $\tilde\phi(\varepsilon,\tilde x)$ for $\varepsilon \in I$, $\tilde x \in \tilde L$, we have

$$\mathrm{supp}\, T_{\mu^{-1}x} S_\varepsilon \tilde\phi(\varepsilon,\mu^{-1}x) \subseteq \overline B_{r\varepsilon}(\mu^{-1}x) \subseteq \tilde M$$

for $\varepsilon \leq \eta$, $x \in L$. Applying $\bar\mu^J$ we obtain, by (7.3),

$$\mathrm{supp}\, \mathrm{pr}_1 \bar\mu^J T S^{(\varepsilon)}(\tilde\phi(\varepsilon,\mu^{-1}x),\mu^{-1}x) \subseteq \overline B_{lr\varepsilon}(x) \cap M$$

and, finally, $\mathrm{supp}\,\phi(\varepsilon,x) \subseteq \overline B_{lr}(0)$ for $\varepsilon \leq \eta$, $x \in L$. It follows that for each $\alpha \in \mathbb{N}_0^s$, $\mathrm{supp}\,\partial^\alpha\phi(\varepsilon,x) \subseteq \overline B_{lr}(0)$ for $\varepsilon < \eta$, $x \in L^\circ$. For the boundedness of $\{\partial^\alpha\phi(\varepsilon,x) \mid (\varepsilon,x) \in (0,\eta_1) \times L^\circ\}$ where $\eta_1 := \min(\eta,\frac{h}{rl}) = \min(\frac{\tilde h}{r},\frac{h}{rl})$ it now suffices to show that for each fixed $\beta \in \mathbb{N}_0^s$,

$$\sup\{|\partial_\xi^\beta \partial_x^\alpha \phi(\varepsilon,x)(\xi)| \mid \varepsilon < \eta_1,\ x \in L,\ \|\xi\| \leq lr\}$$

is finite. For $\varepsilon < \eta_1$, $x \in L^\circ$, $\|\xi\| \leq lr$ (observe that $\mathrm{supp}\,\partial_x^\alpha\phi(\varepsilon,x) \subseteq \overline B_{lr}(0) \subseteq \frac{\Omega-x}{\varepsilon}$) $\partial_\xi^\beta \partial_x^\alpha \phi(\varepsilon,x)(\xi)$ is a sum of terms of the form

$$(7.4) \qquad \begin{array}{l} \partial_{\tilde\xi}^{\beta_0}((\partial_{\tilde x}^{\alpha_0}\tilde\phi)(\varepsilon,\mu^{-1}x))\left(\frac{\mu^{-1}(\varepsilon\xi+x)-\mu^{-1}x}{\varepsilon}\right) \\ \cdot g_0(x) \cdot g_1(\varepsilon,x,\xi) \cdot \ldots \cdot g_p(\varepsilon,x,\xi) \cdot \partial_\xi^{\beta'} \partial_x^{\alpha'} |\det D\mu^{-1}(\varepsilon\xi+x)| \end{array}$$

where g_0 is a certain product of derivatives of components of μ^{-1} (hence bounded on L) and each g_j $(1 \leq j \leq p)$ is some derivative of some component of $\frac{\mu^{-1}(\varepsilon\xi+x)-\mu^{-1}x}{\varepsilon}$, i.e.,

$$g_j(\varepsilon,x,\xi) = \partial_\xi^{\beta_j} \partial_x^{\alpha_j}\left(\frac{\mu^{-1}_{i_j}(\varepsilon\xi+x)-\mu^{-1}_{i_j}x}{\varepsilon}\right).$$

For $\varepsilon \leq \eta_1$, $x \in L$, $\|\xi\| \leq lr$, we have $x + \varepsilon\xi \in \overline{B}_h(x) \subseteq M$. Thus the last factor in (7.4) is uniformly bounded, as is $\partial_\xi^{\beta_0}((\partial_{\tilde{x}}^{\alpha_0}\tilde{\phi})(\varepsilon, \mu^{-1}x))(\tilde{\xi})$ for $\varepsilon \leq 1$, $x \in L$, $\tilde{\xi} \in \mathbb{R}^s$. It remains to discuss the boundedness of the factors g_j ($j = 1, \ldots, p$). For the sake of simplicity, we replace α_j, β_j, i_j by α, β, i. Considering first the case $|\beta| > 0$ (say, $\beta_k \geq 1$), the uniform boundedness of

$$\partial_\xi^\beta \partial_x^\alpha \frac{\mu_i^{-1}(\varepsilon\xi + x) - \mu_i^{-1}x}{\varepsilon} = \partial_\xi^{\beta - e_k} \partial_x^\alpha \left((\partial_k \mu_i^{-1})(\varepsilon\xi + x)\right)$$

on $\varepsilon \leq \eta_1$, $x \in L$, $\|\xi\| \leq lr$ is evident. If, on the other hand, $|\beta| = 0$, choose a Lipschitz constant $l_{\alpha,i}$ for $\partial_x^\alpha \mu_i^{-1}$ with respect to L, h (in the same way as l was chosen for μ with respect to $\tilde{L}, \tilde{h}$ as to satisfy (7.3)). It follows that

$$\left|\frac{(\partial_x^\alpha \mu_i^{-1})(\varepsilon\xi + x) - (\partial_x^\alpha \mu_i^{-1})(x)}{\varepsilon}\right| \leq \frac{l_{\alpha,i}\|\varepsilon\xi\|}{\varepsilon} \leq l_{\alpha,i}lr$$

which establishes the uniform boundedness on $\varepsilon \leq \eta_1$, $x \in L$, $\|\xi\| \leq lr$ also of this term. Replacing $\varepsilon_0 = \frac{1}{2}\eta$ by $\varepsilon_0 := \frac{1}{2}\eta_1$, we have shown altogether that ϕ is smooth and each derivative $\partial^\alpha \phi$ is bounded on the open neighborhood $(0, \eta_1) \times L^\circ (\subseteq D)$ of $(0, \varepsilon_0] \times K$, as required for satisfying conditions 1) and 2) of Theorem 10.5, (Z).

Finally, assume that all derivatives $\partial_{\tilde{x}}^\alpha \tilde{\phi}(\varepsilon, \tilde{x})$ have asymptotically vanishing moments of order q on some compact subset $\tilde{L}$ of $\tilde{\Omega}$ ($\alpha \in \mathbb{N}_0^s$). We have to show that for all $\beta \in \mathbb{N}_0^s$ satisfying $|\beta| \leq \left[\frac{q+1}{2}\right]$ and for arbitrary $\alpha \in \mathbb{N}_0^s$,

$$(7.5) \quad \langle \xi^\beta, \partial_x^\alpha \phi(\varepsilon, x)(\xi)\rangle = \partial_x^\alpha \int \left(\frac{\mu(\varepsilon\tilde{\xi} + \tilde{x}) - \mu\tilde{x}}{\varepsilon}\right)^\beta \tilde{\phi}(\varepsilon, \tilde{x})(\tilde{\xi})\, d\tilde{\xi} = O(\varepsilon^{[\frac{q+1}{2}]})$$

uniformly for $x = \mu\tilde{x} \in L$ (or $\tilde{x} = \mu^{-1}x \in \tilde{L}$, respectively). Note that the preceding equation is meaningful since there exists $\varepsilon_0 > 0$ such that all the terms occurring therein are defined for $\varepsilon \leq \varepsilon_0$, $x \in L$, $\xi \in \mathbb{R}^s$, $\tilde{\xi} \in \mathbb{R}^s$ resp. $\tilde{\xi}$ ranging over a fixed compact set containing the supports of all $\tilde{\phi}(\varepsilon, \tilde{x})$ with $\varepsilon \leq \varepsilon_0$, $\tilde{x} \in \tilde{L}$ in its interior.

Let us consider the case $\alpha = 0$ first. Expanding μ into a Taylor series up to order n at $\tilde{x}$ we may write the first factor in the integral as a finite sum of terms of the form
(7.6)
$$\frac{1}{\varepsilon^{|\beta|}} \frac{\partial^{\alpha_1} \mu_{i_1}(\tilde{x} + \eta_{|\alpha_1|}\theta_{i_1}\varepsilon\tilde{\xi})}{\alpha_1!} \tilde{\xi}^{\alpha_1}\varepsilon^{|\alpha_1|} \cdots \frac{\partial^{\alpha_{|\beta|}} \mu_{i_{|\beta|}}(\tilde{x} + \eta_{|\alpha_{|\beta|}|}\theta_{i_{|\beta|}}\varepsilon\tilde{\xi})}{\alpha_{|\beta|}!} \tilde{\xi}^{\alpha_{|\beta|}}\varepsilon^{|\alpha_{|\beta|}|}$$

where $1 \leq |\alpha_j| \leq n+1$, $0 < \theta_i < 1$, $\eta_j = 1$ if $j = n+1$ (i.e., if the respective factor is a remainder term of the Taylor series) and $\eta_j = 0$ otherwise. Observe that in the present context, by $\alpha_1, \ldots, \alpha_{|\beta|}$ we are denoting $|\beta|$ variables taking values in $\mathbb{N}_0^s$, yet *not* components of a single variable $\alpha \in \mathbb{N}_0^s$. Letting $\gamma := \sum_{j=1}^{|\beta|} \alpha_j$, the above expression (7.6) contains a factor $\tilde{\xi}^\gamma \varepsilon^{|\gamma|-|\beta|}$. (Note that since all $|\alpha_j| \geq 1$ we have $|\gamma| \geq |\beta|$.) If $|\gamma| - |\beta| \geq \left[\frac{q+1}{2}\right]$ we are done with that particular term, taking into account that the integral has to be taken over a fixed compact set only. If, on the other hand, $|\gamma| - |\beta| < \left[\frac{q+1}{2}\right]$ and all η_j vanish we may use the assumption on $\tilde{\phi}(\varepsilon, \tilde{x})$ since in this case $|\gamma| < \left[\frac{q+1}{2}\right] + |\beta| \leq q+1$. Finally, if there is at least one η_j nonvanishing then at least one $|\alpha_j| = n+1$, implying $|\gamma| \geq n + |\beta|$. Hence, choosing $n \geq \left[\frac{q+1}{2}\right]$ completes the proof for the case $\alpha = 0$.

7.4. DIFFEOMORPHISM INVARIANCE

To deal with the general case, express the operator ∂_x^α occurring in (7.5) in terms of operators $\partial_{\tilde{x}}^{\alpha'}$, according to the chain rule. Now apart from certain partial derivatives of components of μ^{-1} (which are bounded on L), Leibniz' rule yields a sum of terms which are similar to those considered above, with certain derivatives $\partial_{\tilde{x}}^{\alpha''} \partial^{\alpha_j} \mu_{i_j}$ and $\partial_{\tilde{x}}^{\alpha'''} \tilde{\phi}$ replacing $\partial^{\alpha_j} \mu_{i_j}$ and $\tilde{\phi}$, respectively. The powers of ξ resp. ε remaining unchanged, the same reasoning as above establishes (7.5) for arbitrary $\alpha \in \mathbb{N}_0^s$. □

Note that the conclusion of the preceding theorem is also obtained if, instead of $\tilde{\phi} \in \mathcal{C}_b^\infty(I \times \tilde{\Omega}, \mathcal{A}_0(\mathbb{R}^s))$, $\tilde{\phi}$ is only assumed to satisfy the analogs (for $\tilde{\Omega}$) of conditions 1) and 2) of 7.14. Moreover, $\mathcal{A}_0(\mathbb{R}^s)$ can be replaced by $\mathcal{D}(\mathbb{R}^s)$ throughout.

Now **(T7)** and **(T8)** follow from **(T6)** and **(D5)**, due to the particular form of **(D3)** and **(D4)**: Assuming, for example, R to be moderate, R also satisfies condition (Z) of Theorem 10.5. Given $\tilde{K} \subset\subset \tilde{\Omega}$, $\alpha \in \mathbb{N}_0^s$ and $\tilde{\phi} \in \mathcal{C}_b^\infty(I \times \tilde{\Omega}, \mathcal{A}_0(\mathbb{R}^s))$, define ϕ as in **(T6)**. According to the chain rule,

$$\begin{aligned}
\partial_{\tilde{x}}^\alpha \left((\hat{\mu} R)(S_\varepsilon \tilde{\phi}(\varepsilon, \tilde{x}), \tilde{x}) \right) &= \partial_{\tilde{x}}^\alpha \left(R(S_\varepsilon \phi(\varepsilon, \mu \tilde{x}), \mu \tilde{x}) \right) \\
&= \sum_{\beta:\, |\beta| \leq |\alpha|} \partial_x^\beta (R(S_\varepsilon \phi(\varepsilon, x), x))\Big|_{x=\mu \tilde{x}} \cdot g_\beta(\tilde{x})
\end{aligned}$$

where each function g_β is a certain sum of products of partial derivatives of components of μ, hence bounded on $\tilde{K}$. R satisfying (Z) of Theorem 10.5, it follows that for some $N \in \mathbb{N}$, $\partial_{\tilde{x}}^\alpha \left((\hat{\mu} R)(S_\varepsilon \tilde{\phi}(\varepsilon, \tilde{x}), \tilde{x}) \right) = O(\varepsilon^{-N})$ uniformly on K. This shows that also $\hat{\mu} R$ is moderate. If, on the other hand, R is assumed to be negligible, R even passes the negligibility test on test objects ϕ being of type $[\mathrm{A}_1^\infty]_{K,q}$, according to Theorem 7.9. Now a similar reasoning as in the case of moderateness, this time using Corollary 10.7 in place of Theorem 10.5, establishes the invariance of negligibility under the action of a diffeomorphism. Thus we have shown

THEOREM 7.15. **(T7)** ([**26**], 25.)
$\mathcal{E}_M$ is invariant under $\hat{\mu}$, i.e., $\hat{\mu}^C$ maps $\mathcal{E}_M(\Omega)$ into $\mathcal{E}_M(\tilde{\Omega})$.

THEOREM 7.16. **(T8)** ([**26**], 25.)
$\mathcal{N}$ is invariant under $\hat{\mu}$, i.e., $\hat{\mu}^C$ maps $\mathcal{N}(\Omega)$ into $\mathcal{N}(\tilde{\Omega})$.

Having completed all the steps of the general construction scheme in chapter 3 we finally reach the goal of this chapter, the definition of the algebra itself:

DEFINITION 7.17. **(D6)** ([**26**], 19.)

$$\mathcal{G}(\Omega) := \mathcal{E}_M(\Omega)/\mathcal{N}(\Omega)$$

Since the respective ideals of negligible functions are invariant under $D_i : \mathcal{E}_M(\Omega) \to \mathcal{E}_M(\Omega)$ as well as under $\hat{\mu} : \mathcal{E}_M(\Omega) \to \mathcal{E}_M(\tilde{\Omega})$, both these maps factorize via the respective quotients to yield maps (which we denote by the same symbols) $D_i : \mathcal{G}(\Omega) \to \mathcal{G}(\Omega)$ and $\hat{\mu} : \mathcal{G}(\Omega) \to \mathcal{G}(\tilde{\Omega})$. This completes the (functorial) construction of a differential algebra containing $\mathcal{D}'(\Omega)$ in such a way as to extend the usual product on $\mathcal{C}^\infty(\Omega)$.

If we had decided to perform this construction in the J-frame we would have obtained objects $\mathcal{G}^J(\Omega)$ isomorphic to the $\mathcal{G}^C(\Omega)$ above: Indeed, also T^* factorizes via quotients with respect to $\mathcal{N}^J(\Omega)$ resp. $\mathcal{N}^C(\Omega)$, thereby inducing a bijection between the J- and C-variant of the diffeomorphism invariant Colombeau algebra at hand.

Next, we give an example of a test object $\phi(\varepsilon, x)$ in the sense of **(D3)** and a distribution $u \in \mathcal{D}'(\Omega)$ such that on every strip $((0, \varepsilon_0] \times \Omega) \cap D$, the map $x \mapsto (\iota u)(S_\varepsilon \phi(\varepsilon, x), x)$ is not even locally bounded (hence, *a fortiori*, neither smooth) where $\varepsilon_0 \in I$ is arbitrary and
$$D := \{(\varepsilon, x) \mid (\phi(\varepsilon, x), x) \in U_\varepsilon(\Omega)\} = \{(\varepsilon, x) \mid \operatorname{supp} S_\varepsilon \phi(\varepsilon, x) \subseteq \Omega - x\}$$
is the natural maximal domain of definition of $(\iota u)(S_\varepsilon \phi(\varepsilon, x), x)$. This phenomenon is due to the mismatch of the respective smoothness notions for ιu and ϕ. It cannot occur on sets of the form $(0, \varepsilon(K)] \times K$ where $K \subset\subset \Omega$ and $\varepsilon(K)$ is chosen suitably with respect to K, according to the discussion following **(D4)** (Definition 7.3).

EXAMPLE 7.18. We employ the notation introduced in Example 5.9. In particular, by $x, y; \xi, \eta$ we now denote coordinates of points $(x, y), (\xi, \eta) \in \mathbb{R}^2$, (ξ, η) replacing the former (x, y) due to the need for additional variables x, y in $\phi(x, y)(\xi, \eta)$. Let Ω, ψ, φ, u be defined as in Example 5.9. Choose a smooth non-decreasing function $\nu : \mathbb{R} \to \mathbb{R}$ taking the constant value k on each of the intervals $I_k := [k - \frac{1}{4}, k + \frac{1}{4}]$ ($k \in \mathbb{Z}$), respectively. Define $\phi(\varepsilon, x, y)(\xi, \eta) := \sin \pi y \cdot S_{|\nu(y)|} \psi(\xi, \eta + \nu(y)) + \varphi(\xi, \eta)$. Then, obviously, $\phi \in \mathcal{C}_b^\infty(I \times \Omega, \mathcal{A}_0(\mathbb{R}^s))$ (note that also $|\nu|$ is smooth). Letting $x := 0$, $y \in I_k \setminus \{k\}$, $\varepsilon := \frac{1}{|k|}$ ($0 \neq k \in \mathbb{Z}$), $(\xi, \eta) \in \Omega$, we obtain
$$\begin{aligned}(\iota u)(S_\varepsilon \phi(\varepsilon, x, y), x, y) &= \sin \pi y \cdot \langle u, S_\varepsilon S_{|\nu(y)|} \psi(\xi, \eta + \nu(y) - y)\rangle \\ &= \sin \pi y \cdot \langle u, \psi(\xi, \eta + k - y)\rangle.\end{aligned}$$
Substituting $t := y - k$ in the last expression (note that $0 < |t| \leq \frac{1}{4}$) yields $(-1)^k \sin \pi t \cdot \langle u, \psi(\xi, \eta - t)\rangle$, the modulus of which tends to infinity as $t \to 0$ (i.e., as $y \to k$) according to Example 5.9.

EXAMPLE 7.19. We demonstrate that $\bar{\mu}^J$ in general is not τ_0-τ_0-continuous. To this end, it is sufficient to show that $\Phi_\mu : \varphi \mapsto (\varphi \circ \mu^{-1}) \cdot |(\mu^{-1})'| = (\varphi \circ \mu^{-1}) \cdot \frac{1}{|\mu' \circ \mu^{-1}|}$ is not a continuous map from $\mathcal{A}_0(\tilde{\Omega})$ into $\mathcal{A}_0(\Omega)$ with respect to the topology τ induced by the (LF)-topology of $\mathcal{D}(\mathbb{R})$, for some open subsets $\tilde{\Omega}, \Omega$ of $\mathbb{R}$ and $\mu : \tilde{\Omega} \to \Omega$ a suitable diffeomorphism. Consider $\tilde{\Omega} := \Omega := (0, \infty)$; choose $\rho \in \mathcal{D}(\mathbb{R})$ as in Example 5.9, that is, $\operatorname{supp} \rho \subseteq [0, 2]$, $\int \rho = 0$ and $\rho(x) = \exp(-\frac{1}{x})$ for $0 < x \leq 1$. Further, fix any $\psi \in \mathcal{A}_0(\Omega)$. Then $\varphi_n(\xi) := \frac{1}{n} \rho(\xi - \frac{1}{n}) + \psi(\xi)$ defines a sequence converging to ψ in $\mathcal{A}_0(\Omega)$ with respect to τ. Now consider $\mu : \tilde{\Omega} \to \Omega$ defined by $\mu(\xi) := \frac{\xi}{3} \exp(-\frac{3}{\xi})$. Then the sequence formed by $\Phi_\mu(\varphi_n)$ is not even bounded with respect to τ: Evaluating $\Phi_\mu(\frac{1}{n} \rho(\xi - \frac{1}{n}))$ at $\mu(\frac{2}{n})$ yields
$$\frac{1}{n} \cdot \rho\left(\frac{2}{n} - \frac{1}{n}\right) \cdot \frac{1}{\mu'(\frac{2}{n})} = \frac{1}{n} \cdot e^{\frac{n}{2}} \cdot \frac{6}{2 + 3n} \to \infty \quad (n \to \infty).$$

To conclude this chapter, we briefly introduce the notion of association into the diffeomorphism-invariant setting:

DEFINITION 7.20. $R_1, R_2 \in \mathcal{G}(\Omega)$ are called associated ($R_1 \approx R_2$) if the following condition holds: $\forall \psi \in \mathcal{D}(\Omega) \exists q \in \mathbb{N} \; \forall \phi \in \mathcal{C}_b^\infty(I \times \Omega, \mathcal{A}_q(\mathbb{R}^s))$:
$$\lim_{\varepsilon \to 0} \int (R_1 - R_2)(S_\varepsilon \phi(\varepsilon, x), x) \psi(x) \, dx = 0$$

(where we have used the C-formalism). The concept of associated distribution as well as the basic properties of association are analogous to the non-diffeomorphism invariant case.

CHAPTER 8

Sheaf properties

Localization properties of elements of $\mathcal{G}$ are most conveniently formulated in terms of sheaves. This chapter presents the relevant notations and results.

Let $\Omega \subseteq \Omega'$. Then $U(\Omega') \subseteq U(\Omega)$ and for $R \in \mathcal{G}(\Omega)$ we denote by $R|_{\Omega'}$ the restriction of R to $U(\Omega')$.

THEOREM 8.1. $\Omega \to \mathcal{G}(\Omega)$ *is a fine sheaf of differential algebras.*

PROOF. Let $\Omega = \bigcup_{\alpha \in A} \Omega_\alpha$. We have to show that

(S1) If $R_1, R_2 \in \mathcal{G}(\Omega)$ and $R_1|_{\Omega_\alpha} = R_2|_{\Omega_\alpha}$ for all $\alpha \in A$ then $R_1 = R_2$.

(S2) If for each $\alpha \in A$ we are given $R_\alpha \in \mathcal{G}(\Omega_\alpha)$ such that $R_\alpha|_{\Omega_\alpha \cap \Omega_\beta} = R_\beta|_{\Omega_\alpha \cap \Omega_\beta}$ for all α, β with $\Omega_\alpha \cap \Omega_\beta \neq \emptyset$ then there exists some $R \in \mathcal{G}(\Omega)$ with $R|_{\Omega_\alpha} = R_\alpha$ for all $\alpha \in A$.

(F) If $(\Omega_\alpha)_\alpha$ is locally finite there exists a family of sheaf morphisms $\eta_\alpha : \mathcal{G} \to \mathcal{G}$ such that
  (i) $\sum_{\alpha \in A} \eta_\alpha = \text{id}$.
  (ii) $\eta_\alpha(\mathcal{G}_x) = 0$ for all x in a neighborhood of $\Omega \setminus \Omega_\alpha$ (where $\mathcal{G}_x$ denotes the stalk of $\mathcal{G}$ at x).

Noting that any $K \subset\subset \Omega$ can be written as $K = \bigcup_{\alpha \in A} K_\alpha$, $K_\alpha \subset\subset \Omega_\alpha$, $K_\alpha = \emptyset$ $\forall \alpha \in A \setminus H$, $|H| < \infty$, (S1) follows directly from Definition 7.3. For proving (S2) we adapt a construction from [**26**], 21, Theorem 1. Choose a locally finite covering $(W_j)_{j \in \mathbb{N}}$ of Ω such that for each $j \in \mathbb{N}$ there exists $\alpha(j)$ with $\overline{W}_j \subset\subset \Omega_{\alpha(j)}$. Let $(\chi_j)_{j \in \mathbb{N}}$ be a partition of unity subordinate to $(W_j)_{j \in \mathbb{N}}$. Moreover, for each $j \in \mathbb{N}$ let $\theta_j \in \mathcal{D}(\Omega_{\alpha(j)})$, $\theta_j \equiv 1$ in a neighborhood of $\overline{W}_j$ and let $\psi_j \in \mathcal{A}_0(\Omega_{\alpha(j)})$. The map

$$\pi_j(\varphi, x) := (\theta_j(\,.\,+x)\varphi - (\int \theta_j(\xi)\varphi(\xi - x)d\xi - 1)\psi_j(\,.\,+x), x)$$

is smooth from $U(\Omega)$ to $T^{-1}(\mathcal{A}_0(\Omega_{\alpha(j)}) \times \Omega))$ and $\pi_j|_{U(W_j)} = \text{id}$. Then for each $j \in \mathbb{N}$ $R_j : U(\Omega) \to \mathbb{C}$,

$$R_j(\varphi, x) = \begin{cases} \chi_j(x) R_{\alpha(j)}(\pi_j(\varphi, x)) & x \in \Omega_{\alpha(j)} \\ 0 & x \notin \Omega_{\alpha(j)} \end{cases}$$

is smooth and $R_j|_{W_j} = R_{\alpha(j)}|_{W_j}$. Since $(W_j)_{j \in \mathbb{N}}$ is locally finite

$$R(\varphi, x) := \sum_{j \in \mathbb{N}} R_j(\varphi, x)$$

is an element of $\mathcal{E}(\Omega)$. To show that R is moderate we first note that in a neighborhood of any $K \subset\subset \Omega$ only finitely many R_j do not vanish identically, so it is enough to estimate one single R_j. Let $\phi \in \mathcal{C}_b^\infty(I \times \Omega, \mathcal{A}_0(\mathbb{R}^s))$ and choose $L \subset\subset W_j$ with $\text{supp}(\chi_j) \subset\subset L$. There exists $\varepsilon_0 > 0$ such that for all $\varepsilon \leq \varepsilon_0$ and all x in a compact neighborhood of L in W_j $\mathrm{supp}(S_\varepsilon\phi(\varepsilon,x)) \subseteq W_j - x$, so $(S_\varepsilon\phi(\varepsilon,x),x) \in U(W_j)$. On this set, $\pi_j = \mathrm{id}$ from which the claim follows by our assumption on R_j.

To establish (S2), by (S1) it suffices to show that $R|_{\Omega_\alpha \cap W_k} = R_{\alpha(k)}|_{\Omega_\alpha \cap W_k}$ for all $k \in \mathbb{N}$ and all $\alpha \in A$ (Note that $R_{\alpha(k)}|_{\Omega_\alpha \cap W_k} = R_\alpha|_{\Omega_\alpha \cap W_k}$ for any α by the assumption in (S2)). Now

$$(8.1) \qquad R|_{\Omega_\alpha \cap W_k} - R_{\alpha(k)}|_{\Omega_\alpha \cap W_k} = \sum_{j \neq k} \chi_j (R_{\alpha(j)} \circ \pi_j - R_{\alpha(k)})|_{\Omega_\alpha \cap W_k}$$

For $K \subset\subset \Omega_\alpha \cap W_k$ and $j \neq k$ set $L = K_1 \cap \mathrm{supp}(\chi_j)$. Let $\phi \in \mathcal{C}_b^\infty(I \times (\Omega_\alpha \cap W_k), \mathcal{A}_0(\mathbb{R}^s))$. Then for x in a neighborhood $M \subset\subset \Omega_\alpha \cap W_j \cap W_k$ of L and sufficiently small ε, $(S_\varepsilon\phi(\varepsilon,x),x) \in U(\Omega_\alpha \cap W_j \cap W_k)$, so $\pi_j(S_\varepsilon\phi(\varepsilon,x),x) = (S_\varepsilon\phi(\varepsilon,x),x)$. Hence the $\mathcal{N}$-estimates for the j-th term in (8.1) follow from those of $R_{\alpha(j)}|_{\Omega_{\alpha(j)} \cap \Omega_{\alpha(k)}} - R_{\alpha(k)}|_{\Omega_{\alpha(j)} \cap \Omega_{\alpha(k)}}$ on M and the fact that χ_j vanishes identically in a neighborhood of $K \setminus M^\circ$. Finally, for proving (F) set (for $\Omega' \subseteq \Omega$)

$$\eta_\beta|_{\Omega'} := \mathcal{G}(\Omega') \ni R \to \sum_{\{j | \beta = \alpha(j)\}} \chi_j \left(R|_{\Omega' \cap W_j} \circ \pi_j|_{\Omega'} \right)$$

$\square$

CHAPTER 9

Separating the basic definition from testing

Having introduced a diffeomorphism invariant Colombeau algebra in chapter 7, we briefly return to the general discussion of full Colombeau algebras. Regarding the definitions of moderateness resp. negligibility (**(D3)**,**(D4)**), we have adopted the terminology of "testing" in chapter 3: By Definition **(D1)**, certain "objects" (i.e., functions) R are specified; those which are singled out by Definition **(D3)** as being moderate serve as representatives of elements of the algebra $\mathcal{G} = \mathcal{E}_M/\mathcal{N}$ of generalized functions. The process of deciding whether an object R belongs to $\mathcal{E}_M$ (or to $\mathcal{N}$, respectively) has been called "testing for moderateness resp. negligibility". It is performed by scaling "test objects" of the appropriate type by the operator S_ε (as well as translating them appropriately, whenever the J-formalism is used), plugging them into R and analyzing the resulting behaviour of R on these "paths" as $\varepsilon \to 0$. Depending on the type of Colombeau algebra that is to be constructed, test objects take different forms, for example, φ ([**10**]), $\phi(\varepsilon)$ ([**13**]), $\phi(x)$ ([**26**]) or $\phi(\varepsilon, x)$ ([**26**], [**41**]). As opposed to that, the objects R themselves do *not* depend in any way on ε; neither do they depend on x via the first argument (the "φ-slot"). In other words, R accepts only certain pairs (φ, x) as arguments where $\varphi \in \mathcal{A}_0(\mathbb{R}^s)$, $x \in \mathbb{R}^s$. Summarizing, we adopt the following policy:

Defining the objects $R \in \mathcal{E}(\Omega)$ is to be separated strictly from testing them.

This decision is based on the following reasons: First, it makes the objects simpler and the theory easier to comprehend, yet without restricting its potential. Second, it provides a unifying framework and a common terminology by means of which the different versions of Colombeau algebras and their relations to each other can be analyzed. Finally, it is crucial for the development of algebras of nonlinear generalized functions on smooth manifolds, if this is to be achieved in terms of intrinsic objects; this task is deferred to a subsequent paper ([**24**], jointly with J. Vickers).

Supposing R to be the image of a non-smooth distribution u under the corresponding embedding ι into $\mathcal{E}_M$, $R(S_\varepsilon \phi, x)$ can be thought of as a regularization of u: Indeed, $S_\varepsilon \phi$ tends to the delta distribution weakly, due to $\int \phi \equiv 1$. In this sense, $S_\varepsilon \phi$ (e.g., $S_\varepsilon \phi(\varepsilon, x)$) represents a "smoothing process" in its totality, for all $\varepsilon \in I$ and on the whole x-domain Ω. Separating the definition of the objects R from testing them thus amounts to assuming that R does not respond to the smoothing process as a whole but only to its particular stages (represented by single elements φ of $\mathcal{A}_0(\mathbb{R}^s)$). In the literature, three variants of increasing complexity can be distinguished:

1. The objects R take (certain) pairs (φ, x) as arguments; testing is performed by inserting $(S_\varepsilon \phi(\varepsilon, x), x)$ into R; the behaviour of $R(S_\varepsilon \phi(\varepsilon, x), x)$ has to be studied.
2. Each object R is given by a family $(R_\varepsilon)_{\varepsilon \in I}$ of functions R_ε as in 1.; testing is performed by investigating $(R_\varepsilon(S_\varepsilon \phi(\varepsilon, x), x))_\varepsilon$.
3. The objects R are defined on some set of pairs $(\mathcal{S}, x)$ where $\mathcal{S} = ((\varepsilon, x) \mapsto S_\varepsilon \phi(\varepsilon, x))$ resp. $\mathcal{S} = ((\varepsilon, x) \mapsto \phi(\varepsilon, x))$ represents some "smoothing process". For testing R, $R(\mathcal{S}, x)$ (which, in turn, has to be dependent on ε!) has to be studied as $\varepsilon \to 0$.

The first of the above variants is the one corresponding to "separation of definitions from testing". Any object from level i gives rise to an object of level $(i+1)$ ($i = 1, 2$) by the following assignments:

$$\begin{array}{lll}
\text{level 1} \to \text{level 2} & R_\varepsilon := R & \text{(for all } \varepsilon\text{)} \\
\text{level 2} \to \text{level 3} & (R(\mathcal{S}, x))(\varepsilon, x) := R_\varepsilon(S_\varepsilon \phi(\varepsilon, x), x).
\end{array}$$

Jelínek in [26], Definition 5, definitely chose level 1 for performing his construction: This is made explicit in the last paragraph of item 2 of [26] (see also the discussion in item 3 of [26]). As opposed to that, Definition 5 of [41], e.g., clearly aims at level 3 (the following definition of moderateness (Definition 6), however, is ambiguous since it is not clear in which way $R(\mathcal{S}, x)$ (using our notation) depends on ε). The authors of [13], on the other hand, introduced their basic objects $\mathcal{R} \in \mathcal{E}(\Omega)$ in Definition 2 as smooth maps $\mathcal{R} : \mathcal{A}_0 \times \Omega \to \mathbb{C}^I$ where $\mathcal{A}_0$ denotes a certain set of bounded paths $\phi(\varepsilon)$ (i.e., smoothing processes that are independent of $x \in \Omega$). At first glance, this seems to be a clear indication that it was level 3 they had in mind. In the following line, however, $\mathcal{R}(\phi, x)_\varepsilon$ is specified to be of the form $R(S_\varepsilon \phi(\varepsilon), x)$ (using our notation S_ε for the scaling operator) which has the appearance of level 2, generated by an object R of level 1 in the way described above. As the case may be, using $\mathbb{C}^I$ as range space for $\mathcal{R}$ instead of $\mathbb{C}$ definitely incorporates a certain part of the testing procedure into the definition of the basic objects by introducing ε as parameter from the very beginning.

CHAPTER 10

Characterization results

The aim of this chapter is to derive several characterizations of moderateness and negligibility, respectively, which turned out to be indispensable tools in establishing the diffeomorphism invariance of the algebra constructed in chapter 7. Moreover, these characterizations will serve as a basis for an intrinsic formulation of the theory on manifolds ([**24**]).

We begin by proving Theorem 7.9. To this end, we introduce "descending" sequences of linear projections $P_0, \ldots, P_m$ with the property that P_0 acts as the identity operator on $\mathcal{A}_0(\mathbb{R}^s)$, P_m projects $\mathcal{A}_0(\mathbb{R}^s)$ onto $\mathcal{A}_q(\mathbb{R}^s)$ and the range of P_j is of codimension 1 in the range of P_{j-1} ($j = 1, \ldots, m$).

Fix $q \in \mathbb{N}$ and $r > 0$. Enumerate $\{\beta \mid 1 \leq |\beta| \leq q\}$ in an arbitrary manner as $\{\beta_1, \ldots, \beta_m\}$. Since the family $\{\xi^\beta \mid 0 \leq |\beta| \leq q\}$ is linearly independent in $\mathcal{D}'(B_r(0))$, there exist $\varphi_1, \ldots, \varphi_m \in \mathcal{A}_{00}(\mathbb{R}^s) \cap \mathcal{D}(B_r(0))$ satisfying $\int \xi^{\beta_i} \varphi_j(\xi) \, d\xi = \delta_{ij}$ ($1 \leq i, j \leq m$). Now set

$$P_j := \mathrm{id}_{\mathcal{A}_0(\mathbb{R}^s)} - \sum_{i=1}^{j} \varphi_i \otimes \xi^{\beta_i},$$

that is,

$$P_j(\varphi) := \varphi - \sum_{i=1}^{j} \left(\int \xi^{\beta_i} \varphi(\xi) \, d\xi \right) \cdot \varphi_i$$

for $\varphi \in \mathcal{A}_0(\mathbb{R}^s)$, $j = 0, \ldots, m$. Obviously, the operators P_j satisfy the properties mentioned above. As to using projections as defined above and the mean value theorem in the subsequent proof we follow [**26**].

Proof of Theorem 7.9. It is clear that condition (4^∞) implies the condition specified in Definition 7.3. So let us prove the converse, assuming, in addition, that R is moderate. Let $K \subset\subset \Omega$, $\alpha \in \mathbb{N}_0^s$ and $n \in \mathbb{N}$ be given. In order to derive the required estimates for $\partial^\alpha(R(S_\varepsilon \phi(\varepsilon, x), x))$ (ϕ being a test object of type $[\mathrm{A}_1^\infty]_{K,q}$) we have to provide sets of the form $\mathcal{A}_{0,M_1}(\mathbb{R}^s) \times U \subseteq U_{\varepsilon, M_2}(\Omega)$ ($M_1 \subset\subset \mathbb{R}^s$, $M_2 \subset\subset \Omega$, U an open subset of Ω), the former of which will serve as domain for the operations of calculus to be performed, according to chapter 6. Choose L with $K \subset\subset L \subset\subset \Omega$ and $r > 0$ such that $\mathrm{supp}\,\phi(\varepsilon, x) \subseteq B_r(0)$ for $\varepsilon \in I$, $x \in L$. Then $\mathrm{supp}\,\partial^\beta \phi(\varepsilon, x) \subseteq B_r(0)$ for $\varepsilon \in I$, $x \in K$, $\beta \in \mathbb{N}_0^s$. Let $M_1 := \overline{B_r}(0)$ and pick M_2 satisfying $L \subset\subset M_2 \subset\subset \Omega$. According to Proposition 6.3 there exists $\varepsilon_0 > 0$ such that for $\varepsilon \leq \varepsilon_0$, $U_{\varepsilon, M_2}(\Omega)$ contains $\mathcal{A}_{0,M_1}(\mathbb{R}^s) \times L$. Hence $\mathcal{A}_{0,M_1}(\mathbb{R}^s) \times L^\circ$ is an appropriate domain for our purpose, supposing $\varepsilon \leq \varepsilon_0$ in the sequel.

To obtain a suitable value of q, corresponding to K, α, n fixed above, we proceed as follows:

1. By Theorem 7.12 (note that R was assumed to be moderate), there exists $N \in \mathbb{N}$ such that for every $k = 0, \ldots, |\alpha|$, for every $\beta \in \mathbb{N}_0^s$ with $0 \leq |\beta| \leq$

$|\alpha| + 1$ and for every bounded subset B of $\mathcal{D}(\mathbb{R}^s)$,
$$\partial^\beta \mathrm{d}_1^k (R \circ S^{(\varepsilon)})(\varphi, x)(\psi_1, \ldots, \psi_k) = O(\varepsilon^{-N}) \qquad (\varepsilon \to 0)$$
uniformly for $x \in K$, $\varphi \in B \cap \mathcal{A}_0(\mathbb{R}^s)$, $\psi_1, \ldots, \psi_k \in B \cap \mathcal{A}_{00}(\mathbb{R}^s)$.

2. By Definition 7.3 there exists $q \in \mathbb{N}$ such that for every $\phi \in \mathcal{C}_b^\infty(I \times \Omega, \mathcal{A}_q(\mathbb{R}^s))$,
$$\partial^\alpha (R(S_\varepsilon \phi(\varepsilon, x), x)) = O(\varepsilon^n) \qquad (\varepsilon \to 0)$$
uniformly for $x \in K$.

3. Without loss of generality, we may suppose $q \geq n + N$.

Now let $\phi \in \mathcal{C}_b^\infty(I \times \Omega, \mathcal{A}_0(\mathbb{R}^s))$ be of type $[\mathrm{A}_1^\infty]_{K,q}$. For the values of q and r determined so far, consider m, $\varphi_1 \ldots, \varphi_m$ and the projections $P_0, \ldots, P_m$ introduced above. Defining $c_i \in \mathcal{C}_b^\infty(I \times \Omega, \mathbb{C})$ by $c_i(\varepsilon, x) := \int \xi^{\beta_i} \phi(\varepsilon, x)(\xi) \, d\xi$ ($i = 1, \ldots, m$), there exists a constant $C \geq 1$ such that $|\partial^\gamma c_i(\varepsilon, x)| \leq C \varepsilon^q (\leq C)$ for all $x \in K$, $0 \leq |\gamma| \leq |\alpha|$, $i = 1, \ldots, m$, due to ϕ being of type $[\mathrm{A}_1^\infty]_{K,q}$. In order to benefit from the moderateness of R in form of the differential condition above, there remains an appropriate bounded subset of $\mathcal{D}(\mathbb{R}^s)$ to be specified. To this end, let
$$B := \Gamma \{ \partial^\gamma \phi(\varepsilon, x) \mid x \in K, \ 0 \leq |\gamma| \leq |\alpha|, \ \varepsilon \in I \} + mC \cdot \Gamma \{ \varphi_i \mid i = 1 \ldots, m \}$$
where ΓA denotes the absolutely convex hull of the subset A of an arbitrary linear space. Since $B \cap \mathcal{A}_0(\mathbb{R}^s) \subseteq \mathcal{A}_{0,M_1}(\mathbb{R}^s)$ and $B \cap \mathcal{A}_{00}(\mathbb{R}^s) \subseteq \mathcal{A}_{00,M_1}(\mathbb{R}^s)$, we may safely use differentials of $R \circ S^{(\varepsilon)}$ for $\varepsilon \leq \varepsilon_0$, evaluated at the respective vectors and $x \in K$.

For $j = 0, \ldots, m$, set $\psi_j(\varepsilon, x) := P_j \phi(\varepsilon, x) := \phi(\varepsilon, x) - \sum_{i=1}^j c_i(\varepsilon, x) \varphi_i$. Then, in particular, $\psi_0 = \phi$ and $\psi_m \in \mathcal{C}_b^\infty(I \times \Omega, \mathcal{A}_q(\mathbb{R}^s))$. Restricting our attention to $x \in K$, $\varepsilon \leq \varepsilon_0$, the following statements are easily verified:
$$\partial^\gamma \psi_m(\varepsilon, x) \in \mathcal{A}_{q,M_1}(\mathbb{R}^s) \qquad (0 \leq |\gamma| \leq |\alpha|),$$
$$\partial^\gamma \psi_j(\varepsilon, x) \in B, \ \varphi_j \in B \qquad (j = 0, \ldots, m, \ 0 \leq |\gamma| \leq |\alpha|),$$
and, for every $t \in [0, 1]$, $j = 1, \ldots, m$,
$$\psi_j(\varepsilon, x) + t \cdot c_j(\varepsilon, x) \varphi_j = \phi(\varepsilon, x) - \sum_{i=1}^{j-1} c_i(\varepsilon, x) \varphi_i - (1 - t) \cdot c_j(\varepsilon, x) \varphi_j \in B.$$

By our choice of q, $\sup_{x \in K} |\partial^\alpha (R(S_\varepsilon \psi_m(\varepsilon, x), x))| = O(\varepsilon^n)$ and we are going to show in the sequel that also
$$\sup_{x \in K} |\partial^\alpha (R(S_\varepsilon \phi(\varepsilon, x), x)) - \partial^\alpha (R(S_\varepsilon \psi_m(\varepsilon, x), x))| = O(\varepsilon^n)$$
which will complete the proof of R satisfying (4^∞). For the sake of simplicity, we will omit the argument (ε, x) for the functions ϕ, ψ_j, c_j in the following. According to the chain rule (the proof for the case $\alpha = 0$ is contained trivially in the argument to follow), we obtain
$$\partial^\alpha (R(S_\varepsilon \phi, x)) - \partial^\alpha (R(S_\varepsilon \psi_m, x)) = \sum_{j=1}^m \left[\partial^\alpha (R(S_\varepsilon \psi_{j-1}, x)) - \partial^\alpha (R(S_\varepsilon \psi_j, x)) \right] =$$
$$\sum_{j=1}^m \sum_{\beta, p} \left[(\partial^\beta \mathrm{d}_1^p (R \circ S^{(\varepsilon)}))(\psi_{j-1}, x)(\partial^{\gamma_1} \psi_{j-1}, \ldots, \partial^{\gamma_p} \psi_{j-1}) - \right.$$
$$\left. (\partial^\beta \mathrm{d}_1^p (R \circ S^{(\varepsilon)}))(\psi_j, x)(\partial^{\gamma_1} \psi_j, \ldots, \partial^{\gamma_p} \psi_j) \right]$$

where the second sum extends over certain $\beta, p; \gamma_1, \ldots, \gamma_p$. Obviously it is sufficient to derive the desired estimate for each of the terms in square brackets separately; thus fix $\beta, p; \gamma_1, \ldots, \gamma_p$. Substituting $\partial^{\gamma_i}\psi_{j-1} = \partial^{\gamma_i}\psi_j + \partial^{\gamma_i}c_j\varphi_j$ $(i = 1, \ldots, p)$ and using multilinearity and symmetry of the iterated differential transforms the square-bracket term into the sum of

$$(\partial^\beta \mathrm{d}_1^p(R \circ S^{(\varepsilon)}))(\psi_{j-1}, x)(\partial^{\gamma_1}\psi_j, \ldots, \partial^{\gamma_p}\psi_j) -$$
(10.1)
$$(\partial^\beta \mathrm{d}_1^p(R \circ S^{(\varepsilon)}))(\psi_j, x)(\partial^{\gamma_1}\psi_j, \ldots, \partial^{\gamma_p}\psi_j)$$

and of $2^p - 1$ terms of the form

(10.2) $(\partial^\beta \mathrm{d}_1^p(R \circ S^{(\varepsilon)}))(\psi_{j-1}, x)(\partial^{\gamma_{i_1}}c_j\varphi_j, \ldots, \partial^{\gamma_{i_l}}c_j\varphi_j, \partial^{\gamma_{i_{l+1}}}\psi_j, \ldots, \partial^{\gamma_{i_p}}\psi_j)$

where $\{i_1 \ldots, i_p\} = \{1, \ldots, p\}$ and $1 \leq l \leq p$. Let us consider (10.2) first. Observing that $\sup_{x \in K} |\partial^{\gamma_{i_1}}c_j| \leq C\varepsilon^q$ and that $\psi_{j-1}, \varphi_j, \partial^{\gamma_{i_2}}c_j\varphi_j, \ldots, \partial^{\gamma_{i_l}}c_j\varphi_j, \partial^{\gamma_{i_{l+1}}}\psi_j, \ldots,$
$\partial^{\gamma_{i_p}}\psi_j$ all are members of B (provided $\varepsilon \leq \varepsilon_0$ and $x \in K$) we can make use of the moderateness of R in form of the property derived previously by means of Theorem 7.12 to conclude that

$$\sup_{x \in K} |(\partial^\beta \mathrm{d}_1^p(R \circ S^{(\varepsilon)}))(\psi_{j-1}, x)(\partial^{\gamma_{i_1}}c_j\varphi_j, \ldots, \partial^{\gamma_{i_l}}c_j\varphi_j, \partial^{\gamma_{i_{l+1}}}\psi_j, \ldots, \partial^{\gamma_{i_p}}\psi_j)| =$$
$$O(\varepsilon^q) \cdot O(\varepsilon^{-N}) = O(\varepsilon^{q-N}) \leq O(\varepsilon^n),$$

due to our choice of q. To handle (10.1), on the other hand, we apply the mean value theorem 4.5 to obtain that the value of the term (10.1) is contained in the closed convex hull of the set of all (complex) numbers

(10.3) $\quad (\partial^\beta \mathrm{d}_1^{p+1}(R \circ S^{(\varepsilon)}))(\psi_j + tc_j\varphi_j, x)(\partial^{\gamma_1}\psi_j, \ldots, \partial^{\gamma_p}\psi_j, c_j\varphi_j)$

where $0 < t < 1$, $x \in K$. Taking into account that $\sup_{x \in K} |c_j(\varepsilon, x)| \leq C\varepsilon^q$ and that each of $\psi_j + tc_j\varphi_j, \partial^{\gamma_1}\psi_j, \ldots, \partial^{\gamma_p}\psi_j, \varphi_j$ is a member of B (provided $\varepsilon \leq \varepsilon_0$ and $x \in K$) the modulus of each of the complex numbers given by (10.3) can be estimated by $C'\varepsilon^q\varepsilon^{-N}$ for some positive constant C' being independent of t, x, again by the differential condition derived from the moderateness of R. Consequently,

$$\sup_{x \in K} |(\partial^\beta \mathrm{d}_1^p(R \circ S^{(\varepsilon)}))(\psi_{j-1}, x)(\partial^{\gamma_1}\psi_j, \ldots, \partial^{\gamma_p}\psi_j) -$$
$$(\partial^\beta \mathrm{d}_1^p(R \circ S^{(\varepsilon)}))(\psi_j, x)(\partial^{\gamma_1}\psi_j, \ldots, \partial^{\gamma_p}\psi_j)| = O(\varepsilon^n),$$

thereby concluding the proof. □

The remaining part of this chapter makes available the means allowing to test moderateness resp. negligibility on test objects ϕ which are defined only on certain subsets of $I \times \Omega$. To this end we first have to provide the technical toolkit for manipulating smooth bounded paths.

LEMMA 10.1. (Partition of unity on I) Let $1 > \varepsilon_1 > \varepsilon_2 > \varepsilon_3 > \ldots \to 0$, $\varepsilon_0 = 2$. Then there exist $\lambda_j \in \mathcal{D}(\mathbb{R})$ $(j = 1, 2, \ldots)$ having the following properties:

1) $\operatorname{supp} \lambda_j = [\varepsilon_{j+1}, \varepsilon_{j-1}]$ 2) $\lambda_j(x) > 0$ for $x \in (\varepsilon_{j+1}, \varepsilon_{j-1})$

3) $\sum_{j=1}^{\infty} \lambda_j(x) \equiv 1$ for $x \in I$ 4) $\lambda_j(\varepsilon_j) = 1$ 5) $\lambda_1(x) = 1$ for $x \in [\varepsilon_1, 1]$

PROOF. For $j \in \mathbb{N}$, choose $\lambda_j^\circ \in \mathcal{D}(\mathbb{R})$ such that $\operatorname{supp} \lambda_j^\circ = [\varepsilon_{j+1}, \varepsilon_{j-1}]$, $\lambda_j^\circ > 0$ on $(\varepsilon_{j+1}, \varepsilon_{j-1})$ and $\lambda_0^\circ \in \mathcal{D}(\mathbb{R})$ such that $\operatorname{supp} \lambda_0^\circ = [1,3]$, $\lambda_0^\circ > 0$ on $(1,3)$. Define $\lambda^\circ := \sum_{j=0}^\infty \lambda_j^\circ$ and $\lambda_j(x) := \dfrac{\lambda_j^\circ(x)}{\lambda^\circ(x)}$ for $x \in (0,3)$. Then $\sum_{j=1}^\infty \lambda_j(x) = \sum_{j=0}^\infty \lambda_j(x) \equiv 1$ for $x \in I$ and it is easy to see that also the remaining four conditions are satisfied. □

LEMMA 10.2. *Assume that for each $K \subset\subset \Omega$ there exists $\varepsilon \in I$ such that the pair (ε, K) satisfies a certain property (P) which is stable with respect to decreasing ε and K in the following sense:*

If (ε_1, K_1) satisfies (P) and $\varepsilon_2 \leq \varepsilon_1$, $K_2 \subseteq K_1$, then also (ε_2, K_2) satisfies (P) $(0 < \varepsilon_1, \varepsilon_2 \leq 1$, $K_1, K_2 \subset\subset \Omega)$.
Then there exists $\psi \in \mathcal{C}^\infty(\Omega)$ satisfying $0 < \psi(x) \leq 1$ for each $x \in \Omega$ such that an ε as above which is appropriate for K with respect to (P) can be chosen as $\eta_K := \min_{x \in K} \psi(x)$, i.e., (η_K, K) satisfies (P) for each $K \subset\subset \Omega$.

PROOF. Let K_n be an increasing sequence of compact subsets of Ω satisfying $K_n \subset K_{n+1}^\circ$ which exhausts Ω ($n \in \mathbb{N}$), e.g. $K_n := \{\lambda \in \Omega \mid |\lambda| \leq n \text{ and } \operatorname{dist}(\lambda, \partial\Omega) \geq \frac{1}{n}\}$. Set $K_0 := \emptyset$ and define open sets G_n ($n \in \mathbb{N}$) by $G_1 := \emptyset$, $G_n := K_n^\circ \setminus K_{n-2}$ ($n \geq 2$). Now choose a partition of unity $(\varphi_n)_{n \geq 1}$ subordinate to the open covering $(G_n)_{n \geq 1}$ of Ω. For $n \geq 2$, let ε_n be such that (ε_n, K_n) satisfies (P), $\varepsilon_n \leq \varepsilon_{n-1}$ (we put $\varepsilon_1 := 1$) and define $\psi(x) := \sum_{n=1}^\infty \varepsilon_n \varphi_n(x)$. Then it is clear that ψ is smooth and takes its values in I. To complete the proof, let $\emptyset \neq K \subset\subset \Omega$. Let $N \in \mathbb{N}$ be minimal such that K is contained in K_N° and consider $x \in K \setminus K_{N-1}^\circ$: Since $x \in K_N$, it cannot be an element of G_n for $n \geq N+2$. On the other hand, since $x \notin K_{N-1}^\circ$, it neither can belong to G_n for $n \leq N-1$ (if $N \geq 2$). Altogether, $x \in G_N \cup G_{N+1}$ which results in $\psi(x) = \varepsilon_N \varphi_N(x) + \varepsilon_{N+1} \varphi_{N+1}(x) \in [\varepsilon_{N+1}, \varepsilon_N]$. Therefore, $\eta_K := \min_{x \in K} \psi(x) \leq \min_{x \in K \setminus K_{N-1}^\circ} \psi(x) \leq \varepsilon_N$. So finally from $\eta_K \leq \varepsilon_N$ and $K \subseteq K_N$ we conclude that (η_K, K) satisfies (P). □

The proof of the preceding lemma also gives the following result (just omit K from the argument and consider $x \in K_N \setminus K_{N-1}^\circ$ instead of $x \in K \setminus K_{N-1}^\circ$):

LEMMA 10.3. *Let K_n be an increasing sequence of compact subsets of Ω satisfying $K_n \subset K_{n+1}^\circ$ which exhausts Ω ($n \in \mathbb{N}$) and let $\varepsilon_1 \geq \varepsilon_2 \geq \ldots > 0$ be given. Then there exists $\psi \in \mathcal{C}^\infty(\Omega)$ satisfying $0 < \psi(x) \leq \varepsilon_n$ for $x \in K_n \setminus K_{n-1}^\circ$ ($K_0 := \emptyset$).*

PROPOSITION 10.4. (Extension of bounded paths) *Let $\phi : D \to \mathcal{A}_0(\mathbb{R}^s)$ where $D \subseteq I \times \Omega$ such that for each $K \subset\subset \Omega$ there exists $\varepsilon_0 > 0$ and a subset U of D which is open in $I \times \Omega$ having the following properties:*

1) $(0, \varepsilon_0] \times K \subseteq U (\subseteq D)$ and ϕ is smooth on U;
2) for all $\alpha \in \mathbb{N}_0^s$, $\{\partial^\alpha \phi(\varepsilon, x) \mid 0 < \varepsilon \leq \varepsilon_0,\ x \in K\}$ is bounded in $\mathcal{D}(\mathbb{R}^s)$.

Then there exist a smooth map $\tilde\phi : I \times \Omega \to \mathcal{A}_0(\mathbb{R}^s)$ and $\sigma \in \mathcal{C}^\infty(\Omega)$ ($0 < \sigma(x) \leq 1$ for all $x \in \Omega$) satisfying

1') $\tilde\phi = \phi$ on $\{(\varepsilon, x) \in I \times \Omega \mid \varepsilon \leq \frac{2}{3} \sigma(x)\}$;
2') $\tilde\phi \in \mathcal{C}_b^\infty(I \times \Omega, \mathcal{A}_0(\mathbb{R}^s))$.
3') $(\tilde\phi(\varepsilon, x), x) \in U_\delta(\Omega)$ for all $(\varepsilon, x) \in I \times \Omega$ and $\delta \leq \sigma(x)$.

10. CHARACTERIZATION RESULTS

In particular, conditions 1') and 3') imply that for each $K \subset\subset \Omega$ there exists $\varepsilon_1 := \min_{x \in K} \sigma(x)$ such that $\tilde\phi = \phi$ on an open neighborhood of $(0, \frac{\varepsilon_1}{2}] \times K$ and $(\tilde\phi(\varepsilon, x), x) \in U_\delta(\Omega)$ for all $(\varepsilon, x) \in I \times K$ and $\delta \leq \varepsilon_1$.

PROOF. First we show that without loss of generality it can be assumed that ε_0 occurring in conditions 1) and 2) above also satisfies the following property 3), in addition to 1) and 2):

3) $(\phi(\varepsilon, x), x) \in U_\delta(\Omega)$ *for all* $0 < \varepsilon, \delta \leq \varepsilon_0$ *and* $x \in K$.

In fact, according to Proposition 6.3 there exists $\eta > 0$ such that $(\phi(\varepsilon, x), x) \in U_\delta(\Omega)$ for all $0 < \varepsilon \leq \varepsilon_0$, $x \in K$ and $0 < \delta \leq \eta$ (observe that $\{\phi(\varepsilon, x) \mid 0 < \varepsilon \leq \varepsilon_0, \ x \in K\}$ is bounded by 2)). Replacing both ε_0 and η by $\min(\varepsilon_0, \eta)$, we see that 1)–3) can be assumed to hold simultaneously.

Now let us say that (ε_0, K) satisfies (P) if 1)–3) are valid for this particular pair (ε_0, K). Then Lemma 10.2 can be applied and provides a function $\sigma \in \mathcal{C}^\infty(\Omega)$ satisfying $0 < \sigma(x) \leq 1$ for each $x \in \Omega$ such that 1)–3) hold for $\min_{x \in K} \sigma(x)$ in place of ε_0. Now let λ_1 be smooth on $\mathbb{R}$, $0 \leq \lambda_1 \leq 1$ and $\lambda_1 \equiv 1$ on $(-\infty, \frac{2}{3}]$, $\lambda_1 \equiv 0$ on $[\frac{5}{6}, +\infty)$. Set $\lambda_2 := 1 - \lambda_1$ and define

$$\tilde\phi(\varepsilon, x) := \lambda_1\left(\frac{\varepsilon}{\sigma(x)}\right) \cdot \phi(\varepsilon, x) + \lambda_2\left(\frac{\varepsilon}{\sigma(x)}\right) \cdot \phi(\sigma(x), x).$$

$\tilde\phi$ is defined on $I \times \Omega$ since the formula above actually involves only values of ϕ on pairs (ε, x) satisfying $\varepsilon \leq \sigma(x)$ and $(0, \sigma(x)] \times \{x\} \subseteq D$ by setting $K := \{x\}$ in 1). In order to show that $\tilde\phi$ is smooth we start by observing that ϕ is smooth on some open neighborhood $U (\subseteq D)$ of $D_\sigma := \{(\varepsilon, x) \mid x \in \Omega, \ 0 < \varepsilon \leq \sigma(x)\}$: Setting $K := \{x\}$ in 1) once more yields an open neighborhood U_x of $(0, \sigma(x)] \times \{x\}$ on which ϕ is smooth. It suffices to take $U := \bigcup_{x \in \Omega} U_x$. Now, since $x \mapsto (\sigma(x), x)$ is a smooth map from Ω into U, $\phi(\sigma(x), x)$ is smooth as a function of x. Taking into account that $\operatorname{supp} \lambda_1(\frac{\varepsilon}{\sigma(x)})$ is a subset of $\{(\varepsilon, x) \mid \varepsilon \leq \frac{5}{6}\sigma(x)\}$ and noting that $\sigma(x) > 0$ for all x we see that also the first term in the definition of $\tilde\phi$ and, hence, also $\tilde\phi$ itself are smooth.

Obviously, $\tilde\phi(\varepsilon, x) = \phi(\varepsilon, x)$ for $\varepsilon \leq \frac{2}{3}\sigma(x)$. Thus 1') is proved.

To show 2'), we have to consider derivatives of $\tilde\phi$ with respect to x on sets of the form $I \times K$ where $K \subset\subset \Omega$ is given. Again we set $\varepsilon_1 := \min_{x \in K} \sigma(x)$. First, observe that on $(0, \varepsilon_1] \times K$ all derivatives $\partial^\beta \phi(\varepsilon, x)$ are bounded by 2). Since they are clearly also bounded on the compact set $\{(\varepsilon, x) \mid x \in K, \ \varepsilon_1 \leq \varepsilon \leq \sigma(x)\}$, they are bounded on the whole of $K_\sigma := \{(\varepsilon, x) \mid x \in K, \ 0 \leq \varepsilon \leq \sigma(x)\}$. Now fix $\alpha \in \mathbb{N}_0^s$; to discuss $\partial^\alpha \tilde\phi$ in detail, we set

$$\tilde\phi_1(\varepsilon, x) := \lambda_1\left(\frac{\varepsilon}{\sigma(x)}\right) \cdot \phi(\varepsilon, x) \quad \text{resp.} \quad \tilde\phi_2(\varepsilon, x) := \lambda_2\left(\frac{\varepsilon}{\sigma(x)}\right) \cdot \phi(\sigma(x), x).$$

Now

$$\partial^\alpha \tilde\phi_1(\varepsilon, x) = \sum_{\beta + \gamma = \alpha} \binom{\alpha}{\beta} \partial^\beta \lambda_1\left(\frac{\varepsilon}{\sigma(x)}\right) \cdot \partial^\gamma \phi(\varepsilon, x).$$

As we have seen above, all derivatives $\partial^\gamma \phi(\varepsilon, x)$ are bounded on K_σ. Expanding $\partial^\beta \lambda_1(\frac{\varepsilon}{\sigma(x)})$ according to the chain rule gives a finite number of terms of the form

$$\lambda_1^{(l)}\left(\frac{\varepsilon}{\sigma(x)}\right) \cdot \varepsilon^l \cdot \partial^{\gamma_1}\left(\frac{1}{\sigma(x)}\right) \cdot \ldots \cdot \partial^{\gamma_l}\left(\frac{1}{\sigma(x)}\right)$$

where $1 \leq l \leq |\beta|$ and $\gamma_1, \ldots, \gamma_l \in \mathbb{N}_0^s$ satisfy $\sum_{i=1}^{l} |\gamma_i| = |\beta|$. Each of these terms is bounded on $I \times K$. Taking into account that all $\partial^\beta \lambda_1(\frac{\varepsilon}{\sigma(x)})$ vanish for $\varepsilon \geq \frac{5}{6}\sigma(x)$ and that K_σ is characterized by $\varepsilon \leq \sigma(x)$, $\partial^\alpha \tilde{\phi}_1$ is bounded on $I \times K$.

The derivatives of $\tilde{\phi}_2$ take the form

$$\partial^\alpha \tilde{\phi}_2(\varepsilon, x) = \sum_{\beta+\gamma=\alpha} \binom{\alpha}{\beta} \partial^\beta \lambda_2\left(\frac{\varepsilon}{\sigma(x)}\right) \cdot \partial^\gamma \phi(\sigma(x), x).$$

The above reasoning showing $\partial^\beta \lambda_1(\frac{\varepsilon}{\sigma(x)})$ to be bounded also applies to $\partial^\beta \lambda_2(\frac{\varepsilon}{\sigma(x)})$. For $\partial^\gamma \phi(\sigma(x), x)$ the chain rule gives a finite sum of terms of the form

(10.4) $$\partial_\varepsilon^k \partial_x^{\gamma_0} \phi(\sigma(x), x) \cdot \partial_x^{\gamma_1} \sigma(x) \cdot \ldots \cdot \partial_x^{\gamma_k} \sigma(x)$$

where $0 \leq k \leq |\gamma|$ and $\gamma_0, \gamma_1, \ldots, \gamma_k \in \mathbb{N}_0^s$ satisfy $\sum_{i=0}^{k} |\gamma_i| = |\gamma|$. Since $\partial_\varepsilon^k \partial_x^{\gamma_0} \phi$ is bounded on the compact subset $\{(\sigma(x), x) \mid x \in K\}$ of U, all the factors in (10.4) are bounded on K. Combining this with the boundedness of $\partial^\beta \lambda_2(\frac{\varepsilon}{\sigma(x)})$ shows that also $\partial^\alpha \tilde{\phi}_2$ (and hence $\partial^\alpha \tilde{\phi}$) is bounded on $I \times K$ which completes the proof of 2'.

Finally, to show 3') let $x \in \Omega$, $\delta \leq \sigma(x)$ and conclude from 3) that in particular $(\phi(\sigma(x), x), x) \in U_\delta(\Omega)$. Now for $\varepsilon \leq \sigma(x)$, both $(\phi(\varepsilon, x), x)$ and $(\phi(\sigma(x), x), x)$ belong to $U_\delta(\Omega)$ and so also $(\tilde{\phi}(\varepsilon, x), x)$ does. On the other hand, for $\sigma(x) < \varepsilon \leq 1$ we have $(\tilde{\phi}(\varepsilon, x), x) = (\phi(\sigma(x), x), x) \in U_\delta(\Omega)$. Therefore, for all $\varepsilon \in I$, $(\tilde{\phi}(\varepsilon, x), x) \in U_\delta(\Omega)$.

The last statement of the proof follows from the fact that $\tilde{\phi}$ and ϕ coincide on the open set $\{(\varepsilon, x) \mid \varepsilon < \frac{2}{3}\sigma(x)\}$ which clearly contains $(0, \frac{\varepsilon_1}{2}] \times K$. □

In the following, we will identify each function $\hat{\phi} : I \to \mathcal{C}^\infty(\Omega, \mathcal{A}_0(\mathbb{R}^s))$ in the natural way with the corresponding function $\phi \in \mathcal{C}^{[\infty, \Omega]}(I \times \Omega, \mathcal{A}_0(\mathbb{R}^s))$ where $\hat{\phi}(\varepsilon)(x) = \phi(\varepsilon, x)$. This identification respects the properties of $\hat{\phi}$ resp. ϕ being smooth (see Theorem 4.2) and/or bounded (in the sense specified in chapter 2).

THEOREM 10.5. (A to Z) *The moderateness of an element R of $\mathcal{C}^\infty(U(\Omega))$ can be tested equivalently on bounded subsets of $\mathcal{C}^\infty(\Omega, \mathcal{A}_0(\mathbb{R}^s))$, on arbitrary bounded paths $\phi : I \to \mathcal{C}^\infty(\Omega, \mathcal{A}_0(\mathbb{R}^s))$ or on paths of the same form depending smoothly on ε (conditions (A), (B), (C), respectively).*

Moreover, equivalent moderateness tests can be performed on larger resp. smaller classes of smooth paths which are distinguished by better resp. poorer properties with respect to the domain of definition of $R(S_\varepsilon \phi(\delta, x), x)$ (conditions (D), (E); (Z)). Formally:

Let $R \in \mathcal{C}^\infty(U(\Omega))$. Then the following conditions are equivalent:

10. CHARACTERIZATION RESULTS

(A)
$$\forall K \subset\subset \Omega \ \forall \alpha \in \mathbb{N}_0^s \ \exists N \in \mathbb{N} \ \forall \mathcal{B} \ \text{(bounded)} \subseteq \mathcal{C}^\infty(\Omega, \mathcal{A}_0(\mathbb{R}^s))$$
$$\exists C > 0 \ \exists \eta > 0 \ \forall \phi \in \mathcal{B} \ \forall \varepsilon \, (0 < \varepsilon < \eta) \ \forall x \in K :$$
$$|\partial^\alpha (R(S_\varepsilon \phi(x), x))| \leq C \varepsilon^{-N}$$

(B)
$$\forall K \subset\subset \Omega \ \forall \alpha \in \mathbb{N}_0^s \ \exists N \in \mathbb{N} \ \forall \phi \in \mathcal{C}_b^{[\infty,\Omega]}(I \times \Omega, \mathcal{A}_0(\mathbb{R}^s))$$
$$\exists C > 0 \ \exists \eta > 0 \ \forall \varepsilon \, (0 < \varepsilon < \eta) \ \forall x \in K :$$
$$|\partial^\alpha (R(S_\varepsilon \phi(\varepsilon, x), x))| \leq C \varepsilon^{-N}$$

(C)
$$\forall K \subset\subset \Omega \ \forall \alpha \in \mathbb{N}_0^s \ \exists N \in \mathbb{N} \ \forall \phi \in \mathcal{C}_b^\infty(I \times \Omega, \mathcal{A}_0(\mathbb{R}^s))$$
$$\exists C > 0 \ \exists \eta > 0 \ \forall \varepsilon \, (0 < \varepsilon < \eta) \ \forall x \in K :$$
$$|\partial^\alpha (R(S_\varepsilon \phi(\varepsilon, x), x))| \leq C \varepsilon^{-N}$$

(D) as condition (C), yet only paths $\phi \in \mathcal{C}_b^\infty(I \times \Omega, \mathcal{A}_0(\mathbb{R}^s))$ are considered such that $(\phi(\varepsilon, x), x) \in U_\delta(\Omega)$ for all $\varepsilon, \delta \in I$ and $x \in K$.

(E) as condition (C), yet only paths $\phi \in \mathcal{C}_b^\infty(I \times \Omega, \mathcal{A}_0(\mathbb{R}^s))$ are considered such that for each $L \subset\subset \Omega$ there exists δ_0 having the the property $(\phi(\varepsilon, x), x) \in U_\delta(\Omega)$ for all $(\varepsilon, x) \in I \times L$ and $\delta \leq \delta_0$.

(Z)
$$\forall K \subset\subset \Omega \ \forall \alpha \in \mathbb{N}_0^s \ \exists N \in \mathbb{N} \ \forall \phi : D \to \mathcal{A}_0(\mathbb{R}^s)) \ (D, \phi \text{ as described below})$$
$$\exists C > 0 \ \exists \eta > 0 \ \forall \varepsilon \, (0 < \varepsilon < \eta) \ \forall x \in K : (\varepsilon, x) \in D \text{ and}$$
$$|\partial^\alpha (R(S_\varepsilon \phi(\varepsilon, x), x))| \leq C \varepsilon^{-N}$$

where $D \subseteq I \times \Omega$ and for D, φ the following holds: For each $L \subset\subset \Omega$ there exists ε_0 and a subset U of D which is open in $I \times \Omega$ such that
1) $(0, \varepsilon_0] \times L \subseteq U(\subseteq D)$ and ϕ is smooth on U;
2) for all $\beta \in \mathbb{N}_0^s$, $\{\partial^\beta \phi(\varepsilon, x) \mid 0 < \varepsilon \leq \varepsilon_0, \ x \in L\}$ is bounded in $\mathcal{D}(\mathbb{R}^s)$.

PROOF. (A) $\Longrightarrow$ (B) is clear since the image of ϕ, the latter being viewed as a function from I into $\mathcal{C}^\infty(\Omega, \mathcal{A}_0(\mathbb{R}^s))$, forms a bounded subset of $\mathcal{C}^\infty(\Omega, \mathcal{A}_0(\mathbb{R}^s))$.

(B) $\Longrightarrow$ (C) is trivial.

(C) $\Longrightarrow$ (A) Assume to the contrary $\neg$(A), i.e.,
$$\exists K \subset\subset \Omega \ \exists \alpha \in \mathbb{N}_0^s \ \forall N \in \mathbb{N} \ \exists \mathcal{B} \ \text{(bounded)} \subseteq \mathcal{C}^\infty(\Omega, \mathcal{A}_0(\mathbb{R}^s))$$
$$\forall C > 0 \ \forall \eta > 0 \ \exists \phi \in \mathcal{B} \ \exists \varepsilon \, (0 < \varepsilon < \eta) \ \exists x \in K:$$
$$|\partial^\alpha (R(S_\varepsilon \phi(x), x))| > C \varepsilon^{-N}$$

Fix K, α as given by $\neg$(A). (C) yields $N \in \mathbb{N}$ with the property described there $(\forall \phi \, (\mathcal{C}^\infty, \text{bounded}) \ldots)$. Now for this N, our assumption $\neg$(A) produces a bounded subset $\mathcal{B}$ of $\mathcal{C}^\infty(\Omega, \mathcal{A}_0(\mathbb{R}^s))$ on which R behaves "badly" in the sense that we can inductively define sequences $\varepsilon_j \in I$, $\phi_j \in \mathcal{B}$ and $x_j \in K$ from which a path $\phi_0(\varepsilon, x)$ can be constructed giving a contradiction to (C) if R is tested on it. Explicitly:

Set $C := 1$, $\eta := 1$ and conclude from $\neg$(A):

$$\exists \varepsilon_1 < 1, \qquad \phi_1 \in \mathcal{B}, \ x_1 \in K : \left|\partial^\alpha (R(S_{\varepsilon_1} \phi_1(x), x))\right|_{x=x_1} \bigg| > 1 \cdot \varepsilon_1^{-N}.$$

Set $C := 2$, $\eta := \min(\frac{1}{2}, \varepsilon_1)$ and conclude from ¬(A):
$$\exists \varepsilon_2 < \min(\frac{1}{2}, \varepsilon_1),\ \phi_2 \in \mathcal{B},\ x_2 \in K : \left|\partial^\alpha(R(S_{\varepsilon_2}\phi_2(x), x))\right|_{x=x_2} \left| > 2 \cdot \varepsilon_2^{-N}.\right.$$
Continuing this way, we get $\varepsilon_0 := 1 > \varepsilon_1 > \varepsilon_2 > \ldots \to 0$, $\phi_j \in \mathcal{B}$, $x_j \in K$ satisfying
$$\left|\partial^\alpha(R(S_{\varepsilon_j}\phi_j(x), x))\right|_{x=x_j} \left| > j \cdot \varepsilon_j^{-N}.\right.$$
Take a partition of unity $(\lambda_j)_j$ on I as provided by Lemma 10.1 and define $\phi_0(\varepsilon, x)$
$$:= \sum_{j=1}^{\infty} \lambda_j(\varepsilon)\phi_j(x) \quad (\varepsilon \in I, x \in \Omega).$$ Clearly, ϕ_0 is smooth from I into $\mathcal{C}^\infty(\Omega, \mathcal{A}_0(\mathbb{R}^s))$
and its image is bounded since it is contained in the convex hull of $\mathcal{B}$. By construction, we get
$$\left|\partial^\alpha(R(S_{\varepsilon_j}\phi_0(\varepsilon_j, x), x))\right|_{x=x_j} \left| = \left|\partial^\alpha(R(S_{\varepsilon_j}\phi_j(x), x))\right|_{x=x_j} \right| > j \cdot \varepsilon_j^{-N},\right.$$
contradicting $\partial^\alpha(R(S_\varepsilon \phi_0(\varepsilon, x), x)) = O(\varepsilon^{-N})$ as required by (C).

(C) $\Longrightarrow$ (D) is trivial.

(D) $\Longrightarrow$ (E) Let K, α be given and choose N by (D). Consider a path $\phi(\varepsilon, x)$ and a constant $\delta_0 > 0$ appropriate for K, both as specified in (E). Define
$$\tilde{\phi}(\varepsilon, x) := S_{\delta_0}\phi(\delta_0 \varepsilon, x) \qquad (\varepsilon \in I,\ x \in \Omega).$$
Since $S_\delta \tilde{\phi}(\varepsilon, x) = S_{\delta\delta_0}\phi(\delta_0\varepsilon, x)$, $(\tilde{\phi}(\varepsilon, x), x) \in U_\delta(\Omega)$ for all $\varepsilon, \delta \in I$ and $x \in K$. (D) now gives $|\partial^\alpha(R(S_\varepsilon \tilde{\phi}(\varepsilon, x), x))| = O(\varepsilon^{-N})$ uniformly on K. However, $\partial^\alpha(R(S_\varepsilon \tilde{\phi}(\varepsilon, x), x)) = \partial^\alpha(R(S_{\delta_0 \varepsilon}\phi(\delta_0 \varepsilon, x), x))$ which implies
$$|\partial^\alpha(R(S_\varepsilon \phi(\varepsilon, x), x))| = O\left(\left(\frac{\varepsilon}{\delta_0}\right)^{-N}\right) = O(\varepsilon^{-N}).$$

(E) $\Longrightarrow$ (Z) Again let K, α be given; this time, of course, choose N according to (E). By Proposition 10.4, to a given path $\phi(\varepsilon, x)$ as in (Z) there exists a bounded path $\tilde{\phi}(\varepsilon, x)$ satisfying the condition in (E) such that $\tilde{\phi}$ and ϕ coincide on an open neighborhood of $(0, \frac{\varepsilon_1}{2}] \times K$, where ε_1 is some positive constant. Therefore
$$|\partial^\alpha(R(S_\varepsilon \phi(\varepsilon, x), x))| = |\partial^\alpha(R(S_\varepsilon \tilde{\phi}(\varepsilon, x), x))| = O(\varepsilon^{-N}).$$

(Z) $\Longrightarrow$ (C) Setting $D := I \times \Omega$ turns a path in the sense of (C) in a path as required for the application of (Z). $\square$

There is an analog to the preceding theorem giving rise to equivalent definitions of negligibility. Observe, however, that each of the conditions in the following theorem is equivalent to the condition in Definition 7.3 even *without* assuming R to be moderate. This latter property has to be assumed in addition to obtain the definition of negligibility.

THEOREM 10.6. (A' to Z') *In properties (A)–(Z) of Theorem 10.5, insert "$\forall n \in \mathbb{N}$" after "$\forall \alpha \in \mathbb{N}_0^s$" and replace "$\exists N \in \mathbb{N}$" by "$\exists q \in \mathbb{N}$", "$\mathcal{A}_0(\mathbb{R}^s)$" by "$\mathcal{A}_q(\mathbb{R}^s)$" and "$C\varepsilon^{-N}$" by "$C\varepsilon^n$", throughout. Then the resulting six conditions (A')–(Z') are mutually equivalent. If R, in addition, is supposed to be moderate, each of them is equivalent to the negligibility of R.*

The proof of the preceding theorem is analogous to that of Theorem 10.5. Finally, in the proof of **(T8)** in chapter 7, we have made use of the following variant of (a part of) 10.6:

COROLLARY 10.7. *In each of conditions* (C′) *and* (Z′) *of Theorem 10.6, replace* "$\mathcal{A}_q(\mathbb{R}^s)$" *by* "$\mathcal{A}_0(\mathbb{R}^s)$" *and consider only test objects* ϕ *all whose derivatives* $\partial_x^\alpha \phi$ *($\alpha \in \mathbb{N}_0^s$) have asymptotically vanishing moments of order q on the compact set K at hand. Then the resulting conditions* (C″) *and* (Z″) *are equivalent.*

PROOF. (Z″) $\Rightarrow$ (C″) being trivial, we have to show the reverse implication. To this end, note that for the property of a test object to have asymptotically vanishing moments on some $K \subset\subset \Omega$, only sets of the form $(0, \varepsilon_0] \times K$ (for some $0 < \varepsilon_0 \leq 1$) are relevant. Yet it is part of the statement of Proposition 10.4 that the extended path $\tilde\phi$ agrees with the given path ϕ on sets of this form, for every given $K \subset\subset \Omega$. Thus the property of having asymptotically vanishing moments is preserved by the extension process $\phi \mapsto \tilde\phi$. Now an argument analogous to that used to prove (C)$\Rightarrow$(Z) in Theorem 10.5 establishes (C″)$\Rightarrow$(Z″). □

Conditions (A)–(C) in Theorem 10.5 are due to J. Jelínek ([**26**], the Remark following Definition 8; the proof of equivalence is only indicated there). The equivalence of condition (Z) with (A)–(C) has to be considered as the technical cornerstone of the diffeomorphism invariance of the Colombeau algebra constructed in chapter 7. Apart from that, it can be of advantage in certain situations (for example, when dealing with applications) not to have to bother too much about the domains of definition of $R(S_\varepsilon \phi(\delta, x), x)$ being too small. On the other hand, it can be useful to have guaranteed a certain minimum size of these domains; for this reason (D) and (E) have been included in the theorem. Last, but not least we felt the need to give precise meaning to statements like "only $\varepsilon > 0$ small enough is relevant" ([**13**], p. 361) or "... in the case when the maps Φ_ε [...] are not defined on the same set. We only need that [...] for all $\varepsilon > 0$ sufficiently small." ([**26**], item 7). In our view, the considerable technical expense required for establishing the results of this chapter clearly shows the necessity of a rigorous treatment to be given to these matters.

CHAPTER 11

Differential Equations

The main application of Colombeau algebras so far has been in the field of differential equations. It is therefore of considerable interest to explore how the changes in the construction of the algebra necessary to ensure diffeomorphism invariance affect the process of solving differential equations in $\mathcal{G}$. To illustrate these changes, in the following we are going to discuss two prototypical examples.

EXAMPLE 11.1. Consider the initial value problem
$$\ddot{x}(t) = f(x(t))\delta(t)$$
(11.1)
$$x(-1) = x_0$$
$$\dot{x}(-1) = \dot{x}_0$$

($f : \mathbb{R} \to \mathbb{R}$ smooth) in $\mathcal{G}(\mathbb{R})$. Equations of type (11.1) arise, e.g., in geodesic equations in impulsive gravitational waves, cf. [30]. For simplicity, we assume that the initial values x_0 and $\dot{x}_0$ are real numbers.

As in the case of non-diffeomorphism invariant Colombeau theory, solving (11.1) requires existence and uniqueness results for the classical equation with $\delta(t)$ replaced by $\varphi(-t)$ (recall that a representative of $\iota(\delta)$ is given by $(\varphi, t) \to \varphi(-t)$). By a standard fixed point argument (cf. [30], Lemma 1) this initial value problem has unique global solutions provided supp(φ) is contained in a sufficiently small neighborhood U (depending on f and the initial conditions) of 0. For $\varphi \in \mathcal{D}(U)$ let $R_1(\varphi, .)$ be this unique solution. Choose some $\chi \in \mathcal{D}(U)$, $\chi \equiv 1$ in a neighborhood V of 0. We claim that $R : (\varphi, t) \to R_1(\chi\varphi, t)$ is a representative of a locally bounded solution to (11.1) and that this solution is unique. (Note that independence of cl[R] from χ follows already from the fact that $S_\varepsilon \varphi \chi = S_\varepsilon \varphi$ for any φ and ε sufficiently small). First of all, in order to show that R is smooth on $(U(\mathbb{R}), \tau_2)$ it obviously suffices to establish smoothness of R_1 on that space. To this end, let $s \to (\varphi_s, t_s)$ be a τ_1-smooth curve into some $\mathcal{A}_{0,H}(\mathbb{R}) \times V \subseteq U_N(\mathbb{R})$ as in Theorem 6.6. Then smoothness of solutions of ODEs with respect to a real parameter at once shows that $s \to R_1(\varphi_s, t_s)$ is C^∞, from which the claim follows by definition of smoothness and Theorem 6.6. (We feel that the ease of this kind of argument is a decisive advantage of calculus in convenient vector spaces as used here compared to earlier approaches to differential calculus in locally convex spaces.)

In order to show that $R \in \mathcal{E}_M(\mathbb{R})$, note that for ε sufficiently small

$$R(S_\varepsilon \varphi, t) = x_0 + \dot{x}_0(t+1) + \int_{-1}^{t} \int_{-1}^{s} f(R(S_\varepsilon \varphi, r)) S_\varepsilon \varphi(-r) dr ds .$$

The key to proving the desired estimates is the characterization of moderateness given in Theorem 7.12 on the one hand and Remark 6.7 on the other: For φ, ψ varying bounded subsets of $\mathcal{A}_0(\mathbb{R})$ (resp. $\mathcal{A}_{00}(\mathbb{R})$) and ε sufficiently small, iterated differentials of $R \circ S^{(\varepsilon)}$ are well defined and can be calculated according to the usual differentiable structure of $\mathcal{A}_0(\mathbb{R}) \times \mathbb{R}$. In particular, differentiation with respect to φ can be interchanged with integration (see the proof of 4.6), and the chain rule gives, e.g.:

$$d_1(R(S_\varepsilon\varphi,t))[\psi] = \int_{-1}^{t}\int_{-1}^{s} f'(R(S_\varepsilon\varphi,r))d_1(R(S_\varepsilon\varphi,r))[\psi]S_\varepsilon\varphi(-r)drds$$

$$+ \int_{-1}^{t}\int_{-1}^{s} f(R(S_\varepsilon\varphi,r))S_\varepsilon\psi(-r)drds$$

so the result follows (using Gronwall's inequality) by induction. Even in this rather simple example it is quite obvious that to check the moderateness condition 7.2 directly would be extremely tedious (resp. unmanageable for more complicated equations). Moreover, uniqueness can be established similarly without even having to perform any differentiations owing to the remark following Theorem 7.13.

EXAMPLE 11.2. The semilinear wave equation
$$(11.2) \qquad \begin{aligned} (\partial_t^2 - \Delta)u &= F(u) + H \\ u|_{\{t<0\}} &= 0 \end{aligned}$$

with $F \in \mathcal{O}_M(\mathbb{R})$ globally Lipschitz, $F(0) = 0$ and $H \in \mathcal{G}$, supp $H \subseteq \{t \geq 0\}$ has been treated in the Colombeau framework in [**34**]. Therefore, we will only indicate those modifications that allow to carry over the existence and uniqueness results achieved there into the current setting. Also, we only treat space dimension 3. Let R_H^η be a representative of H supported in $\{t > -\eta\}$ ($\eta > 0$). For each $\varphi \in \mathcal{A}_0(\mathbb{R}^4)$ let $(x,t) \to R(\varphi,x,t)$ be the smooth solution to
$$\begin{aligned} (\partial_t^2 - \Delta)u(x,t) &= F(u(x,t)) + R_H^\eta(\varphi,x,t) \\ u|_{\{t\leq-\eta\}} &= 0 \end{aligned}$$
Then from Kirchhoff's formula we obtain for $t \geq -\eta$

$$(11.3) \quad R(\varphi,x,t) = \frac{1}{4\pi}\int_{-\eta}^{t}\frac{1}{t-s}\int_{|x-y|=t-s}(F(R(\varphi,y,s)) + R_H^\eta(\varphi,y,s))d\sigma(y)ds$$

Composing R in this formula with a smooth curve as in 11.1, smooth dependence of solutions of Volterra integral equations on real parameters implies smoothness of R in φ and as in the classical case, smooth extension to $t < -\eta$ is possible since $F(0) = 0$. Again using Remark 6.7, φ-differentiation (for $R \circ S^\varepsilon$, ε small, φ, ψ_i (as in Theorem 7.12) varying in bounded sets) interchanges with integration in (11.3). Hence derivation of the necessary $\mathcal{E}_M$- resp. $\mathcal{N}$-estimates for existence resp. uniqueness of solutions is carried out analogously to the classical case, again due to Theorem 7.12 and the remark following Theorem 7.13.

Part 2

On the Foundations of Nonlinear Generalized Functions II

CHAPTER 12

Introduction to Part 2

In Part 2 of this monograph we continue the study of diffeomorphism invariant Colombeau algebras. We will use freely notations and results from Part 1; Also, numbering of chapters, theorems and formulas will be continued.

The main result of chapter 13 permits one to simplify the definition of the ideal $\mathcal{N}$ considerably: It dispenses with taking into account the derivatives of the representative being tested. This applies to virtually all versions of Colombeau algebras. This seemingly technical modification, however, has decisive effects on applications: For example, it makes it considerably easier to prove uniqueness of the solutions of many differential equations. Chapter 14 complements chapter 4 ("Calculus") of Part I by certain results needed in chapter 15. In particular, it is shown that $\mathcal{C}^\infty(U,F)$ is complete with respect to the topology of uniform convergence (on a suitable family of bounded sets) in all derivatives resp. differentials, provided F is complete as a locally convex space. In chapter 15 we show that the diffeomorphism invariant algebra $\mathcal{G}^d(\Omega)$ presented in chapter 7 is not injectively included in the Colombeau algebra $\mathcal{G}^e(\Omega)$ of [10] (which, to be sure, is the standard version among those being independent from the choice of a particular approximation of the delta distribution) by constructing two counterexamples. In chapter 16 we develop a framework allowing to classify the range of algebras which can be positioned between $\mathcal{G}^d(\Omega)$ and (the smooth version of) $\mathcal{G}^e(\Omega)$. In particular, we are going to discuss to which extent at least the definition of the algebra introduced by J. F. Colombeau and A. Meril in [13] has to be modified to obtain diffeomorphism invariance. This leads to the construction of the (diffeomorphism invariant) Colombeau algebra $\mathcal{G}^2(\Omega)$ which is closer to the algebra of [13] than the algebra $\mathcal{G}^d(\Omega)$ (chapter 17). Certain classification results of chapters 16 and 17 are essential for obtaining an intrinsic description of Colombeau algebras on manifolds (see [24]). The concluding chapter 18 points out that also weaker invariance properties than with respect to all diffeomorphisms should be envisaged for Colombeau algebras, in particular regarding applications.

In the following, we will abbreviate $R \circ S^{(\varepsilon)}$ as R_ε, throughout. Terms of the form $\partial^\alpha \mathrm{d}_1^k R_\varepsilon$ always are to be read as $\partial^\alpha \mathrm{d}_1^k(R_\varepsilon)$.

CHAPTER 13

A simple condition equivalent to negligibility

The property of a representative $R \in \mathcal{E}(\Omega)$ of a generalized function $[R] \in \mathcal{G}(\Omega)$ to belong to the ideal $\mathcal{N}(\Omega)$ was defined in 7.3. Theorem 18 ($2°$) of [**26**] resp. Theorem 7.13 of chapter 7 give an equivalent condition replacing the term $\partial^\alpha(R(S_\varepsilon\phi(\varepsilon,x),x))$ occurring in 7.3 by $(\partial^\alpha \mathrm{d}_1^k R_\varepsilon)(\varphi,x)(\psi_1,\ldots,\psi_k)$. Moreover, Theorem 18 ($1°$) of [**26**] shows that we still get a condition equivalent to $R \in \mathcal{N}(\Omega)$ if we simply omit the differential with respect to the first variable φ from the statement of ($2°$), provided R is assumed to be moderate. In the following, we are going to show that a further simplification is possible which might seem rather drastic at first glance: It is not even necessary to consider partial derivatives with respect to $x \in \Omega$. In order to facilitate comparing the conditions mentioned so far we include all of them in the following theorem, though only ($0°$) is new.

THEOREM 13.1. *Let Ω be an open subset of $\mathbb{R}^s$ and $R \in \mathcal{E}_M(\Omega)$. Then each of the following conditions is equivalent to $R \in \mathcal{N}(\Omega)$ (in the sense of 7.3):*

($0°$) $\forall K \subset\subset \Omega \ \forall n \in \mathbb{N} \ \exists q \in \mathbb{N} \ \forall B \ (bounded) \subseteq \mathcal{D}(\mathbb{R}^s)$:
$$R_\varepsilon(\varphi,x) = O(\varepsilon^n) \qquad (\varepsilon \to 0)$$
uniformly for $x \in K$, $\varphi \in B \cap \mathcal{A}_q(\mathbb{R}^s)$.

($1°$) $\forall K \subset\subset \Omega \ \forall \alpha \in \mathbb{N}_0^d \ \forall n \in \mathbb{N} \ \exists q \in \mathbb{N} \ \forall B \ (bounded) \subseteq \mathcal{D}(\mathbb{R}^s)$:
$$\partial^\alpha R_\varepsilon(\varphi,x) = O(\varepsilon^n) \qquad (\varepsilon \to 0)$$
uniformly for $x \in K$, $\varphi \in B \cap \mathcal{A}_q(\mathbb{R}^s)$.

($2°$) $\forall K \subset\subset \Omega \ \forall \alpha \in \mathbb{N}_0^d \ \forall k \in \mathbb{N}_0 \ \forall n \in \mathbb{N} \ \exists q \in \mathbb{N} \ \forall B \ (bounded) \subseteq \mathcal{D}(\mathbb{R}^s)$:
$$\partial^\alpha \mathrm{d}_1^k R_\varepsilon(\varphi,x)(\psi_1,\ldots,\psi_k) = O(\varepsilon^n) \qquad (\varepsilon \to 0)$$
uniformly for $x \in K$, $\varphi \in B \cap \mathcal{A}_q(\mathbb{R}^s)$, $\psi_1,\ldots,\psi_k \in B \cap \mathcal{A}_{q0}(\mathbb{R}^s)$.

PROOF. The equivalence of each of ($1°$) and ($2°$) with $R \in \mathcal{N}(\Omega)$ is a part of Theorem 18 of [**26**]. ($1°$) $\Rightarrow$ ($0°$) being trivial, it remains to show ($0°$) $\Rightarrow$ ($1°$). To this end, we will prove, assuming $R \in \mathcal{E}_M(\Omega)$ to satisfy ($0°$), that R satisfies ($1°$) for $\alpha := e_i$, i.e., $\partial^\alpha = \partial_i$ ($i = 1,\ldots,s$) and that, in addition, $\partial_i R$ again is moderate and satisfies ($0°$). Then it will follow by induction that ($1°$) holds for all $\alpha \in \mathbb{N}_0^s$.

So suppose $R \in \mathcal{E}_M(\Omega)$ to satisfy ($0°$) and let $K \subset\subset \Omega$ and $n \in \mathbb{N}$ be given. For $\delta := \min(1, \mathrm{dist}(K, \partial\Omega))$, set $L := K + \overline{B}_{\frac{\delta}{2}}(0)$. Then $K \subset\subset L \subset\subset \Omega$. Now by moderateness of R and Theorem 7.12, choose $N \in \mathbb{N}$ such that for every bounded subset B of $\mathcal{D}(\mathbb{R}^s)$ the relation $\partial_i^2 R_\varepsilon(\varphi,x) = O(\varepsilon^{-N})$ as $\varepsilon \to 0$ holds, uniformly for $x \in L$, $\varphi \in B \cap \mathcal{A}_0(\mathbb{R}^s)$. Next, by the assumption of ($0°$) to hold for R, choose $q \in \mathbb{N}$ such that, again for every bounded subset B of $\mathcal{D}(\mathbb{R}^s)$, we have $R_\varepsilon(\varphi,x) = O(\varepsilon^{2n+N})$ as $\varepsilon \to 0$, uniformly for $x \in L$, $\varphi \in B \cap \mathcal{A}_q(\mathbb{R}^s)$. Now suppose a bounded subset B of $\mathcal{D}(\mathbb{R}^s)$ to be given; let $\varphi \in B \cap \mathcal{A}_q(\mathbb{R}^s)$, $x \in K$ and $0 < \varepsilon \leq \frac{\delta}{2}$;

hence $x + \varepsilon^{n+N} e_i \in L$. By Taylor's theorem, we conclude (to be precise, separately for the real and imaginary part of R)

$$R_\varepsilon(\varphi, x + \varepsilon^{n+N} e_i) = R_\varepsilon(\varphi, x) + \partial_i R_\varepsilon(\varphi, x) \varepsilon^{n+N} + \frac{1}{2} \partial_i^2 R_\varepsilon(\varphi, x_\theta) \varepsilon^{2n+2N}$$

where $x_\theta = x + \theta \varepsilon^{n+N} e_i$ for some $\theta \in (0,1)$; note that also $x_\theta \in L$. Consequently,

$$\partial_i R_\varepsilon(\varphi, x) = \underbrace{\left(R_\varepsilon(\varphi, x + \varepsilon^{n+N} e_i) - R_\varepsilon(\varphi, x)\right) \varepsilon^{-n-N}}_{O(\varepsilon^{2n+N})} - \underbrace{\frac{1}{2} \partial_i^2 R_\varepsilon(\varphi, x_\theta) \varepsilon^{n+N}}_{O(\varepsilon^{-N})},$$

uniformly for $\varphi \in B \cap \mathcal{A}_q(\mathbb{R}^s)$, $x \in K$. Having demonstrated $\partial_i R_\varepsilon(\varphi, x) = O(\varepsilon^n)$ for all $i = 1, \ldots, s$, observe that $\partial_i(R_\varepsilon) = (\partial_i R)_\varepsilon$. Therefore, $\partial_i R$ again satisfies (0°). According to Theorem 7.10 (which is non-trivial, see the discussion in chapter 7), $\partial_i R$ is also moderate. By the remark made above, this completes the proof. □

The reader acquainted with E. Landau's paper [**31**] will easily recognize the method employed therein to form the basis of the preceding proof (though not mentioned explicitly in [**26**], this equally applies to the proof of (1°) ⇒ (2°) of Theorem 18 of [**26**]).

The part of Theorem 13.1 saying that for moderate functions (the appropriate analog of) condition (0°) is equivalent to negligibility applies to virtually all versions of Colombeau algebras of practical importance, in particular, to the following:

- For the special algebra as defined, e.g., in [**33**], p. 109, just replace the term $R_\varepsilon(\varphi, x)$ in condition (0°) by $u_\varepsilon(x)$.
- For the classical full Colombeau algebra of [**10**] simply drop the uniformity requirement concerning φ from (0°).
- For the diffeomorphism invariant Colombeau algebra $\mathcal{G}^2(\Omega)$ to be introduced in chapter 17, the corresponding result is stated as Theorem 17.9.
- For the special algebra on smooth manifolds the corresponding result follows from the local characterization of generalized functions (see [**39**], 4.4).
- The latter also applies to the intrinsically defined full Colombeau algebra on manifolds ([**24**], Corollary 4.5).

In the first and second of these four instances, the respective proofs are obtained by appropriately slimming down the corresponding argument of the proof of Theorem 13.1.

The seemingly technical difference between (0°) and the remaining conditions (including negligibility of R) has decisive effects on applications: For example, if the uniqueness of a solution of a differential equation is to be shown one supposes R_1, R_2 to be representatives of solutions. Note that this includes the assumption that $R_1, R_2 \in \mathcal{E}_M(\Omega)$, hence Theorem 13.1 may be applied. For $[R_1] = [R_2]$ in $\mathcal{G}(\Omega)$ we have to show that $R := R_1 - R_2 \in \mathcal{N}(\Omega)$. Now it suffices to check condition (0°) rather than (1°) (resp. (2°) resp. the original definition of $R \in \mathcal{N}(\Omega)$), i.e., there is no need to analyze the behaviour of any derivative of R.

Apart from that, condition (0°) is also of theoretical relevance. To give a sample, let us demonstrate that it allows to simplify considerably the proof of statement (iv) of Theorem 7.4 in Part 1 (saying that $(\iota - \sigma)(\mathcal{C}^\infty(\Omega)) \subseteq \mathcal{N}(\Omega)$): Since $\iota f - \sigma f \in \mathcal{E}_M(\Omega)$ by (i) and (ii) of Theorem 7.4, it is sufficient for $\iota f - \sigma f \in \mathcal{N}(\Omega)$

to show that
$$(\iota f - \sigma f)(S_\varepsilon \varphi, x) = \int_{\frac{\Omega-x}{\varepsilon}} [f(z\varepsilon + x) - f(x)]\, \varphi(z)\, dz = O(\varepsilon^{q+1}),$$

uniformly for $x \in K$ and φ ranging over some bounded subset of $\mathcal{A}_q(\mathbb{R}^d)$. This, however, is immediate.

CHAPTER 14

Some more calculus

Both the counterexamples to be constructed in chapter 15 will take the form of infinite series, being absolutely convergent in each derivative. Thus we need a theorem guaranteeing the completeness of $\mathcal{E}(\Omega) = \mathcal{C}^\infty(U(\Omega), \mathbb{C}) \equiv \mathcal{C}^\infty(\mathcal{A}_0(\Omega) \times \Omega, \mathbb{C})$ with respect to the corresponding topology. The remarkable ease of the proof of this generalization of a standard result of elementary real analysis clearly exhibits the virtues of calculus in convenient vector spaces as outlined in chapter 4. To this end, let E, F be locally convex spaces and U an open subset of E. If $f : U \to F$ is smooth, its
n-th differential $\mathrm{d}^n f$ belongs to $\mathcal{C}^\infty(U, L^n(E^n, F))$ where $L^n(E^n, F)$ denotes the space $L(E, \ldots, E; F)$ of n-linear bounded maps from $E \times \cdots \times E$ (n factors) into F. (For $n = 0$, set $L^n(E^n, F) := F$.) On $\mathcal{C}^\infty(U, L^n(E^n, F))$, let τ_{cb}^n denote the topology of uniform F-convergence on subsets of the form $K \times B$ where K is a compact subset of U and B is bounded in $E^n = E \times \cdots \times E$. Let $\mathcal{C}^\infty(U, F)$ carry the initial (locally convex) topology τ^∞ induced by the family $(\mathrm{d}^n, \mathcal{C}^\infty(U, L^n(E^n, F)), \tau_{cb}^n)_{n \geq 0}$, i.e., the topology of uniform convergence of all derivatives (that is to say, differentials) on sets $K \times B$ as above. Note that on $\mathcal{C}^\infty(\mathbb{R}, F)$, τ^∞ is just the usual Fréchet topology of compact convergence in all derivatives.

THEOREM 14.1. *Let E, F be locally convex spaces, assume F to be complete and let U be an open subset of E. Then $\mathcal{C}^\infty(U, F)$ is complete with respect to the topology τ^∞ of uniform F-convergence of all differentials on subsets of the form $K \times B$ where K is a compact subset of U and B is bounded in the appropriate product $E^n = E \times \cdots \times E$. Moreover, for each $p \in \mathbb{N}$, the operator $\mathrm{d}^p : \mathcal{C}^\infty(U, F) \to \mathcal{C}^\infty(U, L^p(E^p, F))$ is continuous if both the domain and the range space carry the respective topology τ^∞.*

PROOF. Let (f_ι) be a net in $\mathcal{C}^\infty(U, F)$ which is Cauchy with respect to τ^∞, that is, suppose $(\mathrm{d}^n f_\iota)$ to be a Cauchy net in $\mathcal{C}^\infty(U, L^n(E^n, F))$ with respect to τ_{cb}^n for each $n = 0, 1, 2, \ldots$. Due to the completeness of F, each net $(\mathrm{d}^n f_\iota)$ has a limit $f^{[n]} : U \times E^n \to F$ with respect to (the obvious extension of) τ_{cb}^n. In particular, (f_ι) converges to some function $f := f^{[0]} : U \to F$. Consider a smooth curve $c : \mathbb{R} \to U$; then for each ι, $f_\iota \circ c$ is smooth from $\mathbb{R}$ to F, its n-th derivative at $t \in \mathbb{R}$ being given as a certain sum of terms of the form $(\mathrm{d}^l f_\iota)(c(t))(c^{(k_1)}(t), \ldots, c^{(k_l)}(t))$ where $1 \leq l \leq n$ and $\sum k_j = n$, due to the chain rule. With t ranging over some compact subset of $\mathbb{R}$, the values attained by $c^{(k)}(t)$ form a compact subset of U resp. E, for each $k \in \mathbb{N}_0$. Now it follows from the Cauchy property of (f_ι) that $(f_\iota \circ c)$ is Cauchy in $\mathcal{C}^\infty(\mathbb{R}, F)$ with respect to uniform convergence in all derivatives on compact sets. From the completeness of the latter space we conclude that the limit of $(f_\iota \circ c)$ exists as a smooth function and is equal to $f \circ c$. This argument being valid for any smooth curve c, f itself is smooth. To establish $f = \lim f_\iota$ with respect to τ^∞, it remains to show that for any $n \in \mathbb{N}$, $\mathrm{d}^n f = f^{[n]}$, i.e., that for all $x \in U$, $v_1, \ldots, v_n \in E$ we have

(14.1) $\qquad (\mathrm{d}^n f)(x)(v_1, \ldots, v_n) = \lim (\mathrm{d}^n f_\iota)(x)(v_1, \ldots, v_n),$

For a straight line $c(t) = x + tv$ we obtain, at $t = 0$, $(g \circ c)^{(n)}(0) = (\mathrm{d}^n g)(x)(v, \ldots, v)$ for any $g \in \mathcal{C}^\infty(U, F)$. Therefore,

$$(\mathrm{d}^n f)(x)(v, \ldots, v) = (f \circ c)^{(n)}(0) = \lim (f_\iota \circ c)^{(n)}(0) = \lim (\mathrm{d}^n f_\iota)(x)(v, \ldots, v).$$

Equation (14.1) now follows by polarization (see, e.g., [28], Lemma 7.13 (1)). Finally, the continuity of d^p with respect to the initial topologies τ^∞ is immediate from the following commutative diagram:

$$\begin{array}{ccc} (\mathcal{C}^\infty(U, F), \tau^\infty) & \xrightarrow{\mathrm{d}^p} & (\mathcal{C}^\infty(U, L^p(E^p, F)), \tau^\infty) \\ \downarrow{\mathrm{d}^{p+n}} & & \downarrow{\mathrm{d}^n} \\ (\mathcal{C}^\infty(U, L^{p+n}(E^{p+n}, F)), \tau_{cb}^{p+n}) & \xrightarrow{\mathrm{id}} & (\mathcal{C}^\infty(U, L^n(E^n, L^p(E^p, F))), \tau_{cb}^n) \end{array}$$

Observe that the lower horizontal arrow is a linear homeomorphism, due to $L^p(E^p, F)$ carrying the topology of uniform convergence on bounded sets. □

For the rest of this chapter, let U denote a (non-empty) open subset of a closed affine subspace E_1 of some locally convex space E, E_0 the linear subspace parallel to E_1 and F a complete locally convex space. *Mutatis mutandis*, Theorem 14.1 is valid also in this slightly more general situation. The vectors $v_1, \ldots, v_n$ to be plugged into $\mathrm{d}^n f(x)$ now have to be taken from E_0, as well as B has to denote a bounded subset of E_0^n.

DEFINITION 14.2. Assume, in addition to the above, that the topology of F is generated by some family $\mathcal{P}$ of semi-norms. For fixed $n \in \mathbb{N}_0$, let $(f_k)_{k \in \mathbb{N}}$ denote a sequence of functions

$$f_k : U \times E_0 \times \cdots \times E_0 \to F \qquad (n \text{ factors } E_0)$$
$$f_k : (x, v_1, \ldots, v_n) \mapsto f_k(x)(v_1, \ldots, v_n).$$

We say that (f_k) is exponentially bounded on $U \times E_0^n$ (an (eb)-sequence, for short) if for each compact subset K of U, each bounded subset B of E_0 and each $p \in \mathcal{P}$ there exists a constant $C(\geq 1)$ such that $p(f_k(x)(v_1, \ldots, v_n)) \leq C^k$ for any $k \in \mathbb{N}$, $x \in K$ and $v_i \in B$ ($i = 1, \ldots, n$).

Define $(f_k) + (g_k) := (f_k + g_k)$ and $\lambda(f_k) := (\lambda f_k)$ ($\lambda \in \mathbb{C}$), as well as $(f_k) \cdot (g_k) := (f_k \cdot g_k)$ provided F is a (complete) locally convex topological algebra. Then the following proposition is immediate, due to $C_1^k + C_2^k \leq (C_1 + C_2)^k$ and $C_1^k C_2^k = (C_1 C_2)^k$:

PROPOSITION 14.3. *The set of (eb)-sequences forms a linear space (resp. an algebra if F is a locally convex algebra with jointly sequentially continuous multiplication) with respect to the operations defined above.*

THEOREM 14.4. *Let $f_k \in \mathcal{C}^\infty(U, F)$ for every $k \in \mathbb{N}$. Assume that for each fixed $n \in \mathbb{N}_0$, $(\mathrm{d}^n f_k)_k$ is (eb) on $U \times E_0^n$. Then $\sum_{k=0}^\infty \frac{1}{k!} f_k$ is convergent with respect to τ^∞ to some $f \in \mathcal{C}^\infty(U, F)$. Moreover, $\mathrm{d}^n f = \sum_{k=0}^\infty \frac{1}{k!} \mathrm{d}^n f_k$ for every $n \in \mathbb{N}_0$ where also the latter series converges with respect to τ^∞.*

PROOF. Fix $n \in \mathbb{N}_0$, a compact subset K of U and a bounded subset B of E_0. Since $(\mathrm{d}^n f_k)_k$ is (eb), $\sum_k \frac{1}{k!} \mathrm{d}^n f_k$ is majorized, uniformly on $K \times B^n$, by $\sum_k \frac{C_n^k}{k!}$ for some constant $C_n (\geq 1)$ depending only on n, K and B. Consequently, $\sum_k \frac{1}{k!} f_k$ is Cauchy with respect to τ^∞. Now both the convergence of $\sum_k \frac{1}{k!} f_k$ and the admissibility of term-wise differentiation follow from Theorem 14.1. □

In the sequel, 14.2–14.4 will only be used for $F = \mathbb{C}$; the extension to locally convex algebras being for free virtually, we chose to state them in the general form to indicate the scope of Theorem 14.4.

CHAPTER 15

Non-injectivity of the canonical homomorphism from $\mathcal{G}^d(\Omega)$ into $\mathcal{G}^e(\Omega)$

For every open subset Ω of $\mathbb{R}^s$, there is a canonical algebra homomorphism Φ from the diffeomorphism invariant Colombeau algebra $\mathcal{G}^d(\Omega)$ of [26] (see chapter 7) to the "classical" (full) Colombeau algebra $\mathcal{G}^e(\Omega)$ introduced in [10], 1.2.2 (the upper subscript e being taken from the title "Elementary Introduction to New Generalized Functions" of the latter monograph). In this chapter, we are going to show that Φ is not injective in general by constructing a representative R of a generalized function $[R] \in \mathcal{G}^d(\Omega)$ such that $[R] \neq 0$, yet $\Phi[R] = 0$.

By superscripts d, e we will distinguish between ingredients (as listed in chapter 3) for constructing $\mathcal{G}^d(\Omega)$ resp. $\mathcal{G}^e(\Omega)$. Observe that superscripts d, e are independent of superscripts J, C as introduced in chapter 5: Each of the (non-isomorphic) algebras $\mathcal{G}^d(\Omega)$, $\mathcal{G}^e(\Omega)$ has equivalent descriptions in the C- and the J-formalism, respectively. As in chapter 7, we will use the C-formalism also in the present context. All the relevant definitions are to be found in chapter 7 (for $\mathcal{G}^d(\Omega)$) resp. [10] (for $\mathcal{G}^e(\Omega)$). For the present purpose, the following of them are of particular importance:

(15.1)
$$\begin{aligned} U^d(\Omega) &:= T^{-1}(\mathcal{A}_0(\Omega) \times \Omega) \\ U^e(\Omega) &:= T^{-1}(\mathcal{A}_1(\Omega) \times \Omega)^1 \end{aligned}$$

$$\begin{aligned} \mathcal{E}^d(\Omega) &:= \mathcal{C}^\infty(U^d(\Omega)) \\ \mathcal{E}^e(\Omega) &:= \{R : U^e(\Omega) \to \mathbb{C} \mid x \mapsto R(\varphi, x) \text{ is smooth on } U_\varphi \text{ for each } \varphi\} \end{aligned}$$

where U_φ denotes the (open) set $\{x \mid (\varphi, x) \in U^e(\Omega)\}$.

From now on, we will omit specifying Ω explicitly whenever it is clear which domain is intended. Let $j : U^e \to U^d$ denote set-theoretic inclusion. To see that the restriction map $\Phi_0 = j^*$ maps $\mathcal{E}^d$ into $\mathcal{E}^e$ we have to pass from C-representatives to J-representatives: Smoothness of $R^d \in \mathcal{E}^d$, by definition, is equivalent to smoothness of $(T^*)^{-1} R^d \in \mathcal{C}^\infty(\mathcal{A}_0(\Omega) \times \Omega)$ while for $R^e \in \mathcal{E}^e$, smoothness of $x \mapsto R^e(\varphi, x)$ is equivalent to smoothness of $x \mapsto (T^*)^{-1} R^e(\varphi(.-x), x)$. From this it is clear that $\Phi_0 R^d \in \mathcal{E}^e$ for $R^d \in \mathcal{E}^d$.

Φ_0 even maps $\mathcal{E}_M^d$ into $\mathcal{E}_M^e$ and $\mathcal{N}^d$ into $\mathcal{N}^e$, respectively. This follows easily by inspecting the corresponding definitions: For $R^d \in \mathcal{E}^d$, $R^e \in \mathcal{E}^e$; we have, by

---

[1]The choice of $\mathcal{A}_1(\Omega)$ rather than $\mathcal{A}_0(\Omega)$ in the definition of $U^e(\Omega)$ is due to Colombeau ([10], 1.2.1). We decided to keep the original form of $\mathcal{G}^e(\Omega)$ although all the results of this chapter would remain valid (and, in fact, even slightly easier to formulate) choosing also $U^e(\Omega)$ to be $T^{-1}(\mathcal{A}_0(\Omega) \times \Omega)$.

definition (omitting the quantifiers "$\forall K \subset\subset \Omega \ \forall \alpha \in \mathbb{N}_0^s \ \exists N \in \mathbb{N}$"),

$$R^d \in \mathcal{E}_M^d \Leftrightarrow \forall \phi \in \mathcal{C}_b^\infty(I \times \Omega, \mathcal{A}_0(\mathbb{R}^s)) : \sup_{x \in K} |\partial^\alpha (R^d(S_\varepsilon \phi(\varepsilon, x), x))| = O(\varepsilon^{-N})$$

$$R^e \in \mathcal{E}_M^e \Leftrightarrow \forall \varphi \in \mathcal{A}_N(\mathbb{R}^s) : \sup_{x \in K} |\partial^\alpha (R^e(S_\varepsilon \varphi, x))| = O(\varepsilon^{-N})$$

Obviously, each test object $\varphi \in \mathcal{A}_N(\mathbb{R}^s)$ can be viewed as a particular case of a test object $\phi \in \mathcal{C}_b^\infty(I \times \Omega, \mathcal{A}_0(\mathbb{R}^s))$ by setting $\phi(\varepsilon, x) := \varphi$ independently of ε, x. Thus from $R^d \in \mathcal{E}_M^d$ it follows that $\Phi_0 R^d \in \mathcal{E}_M^e$. A similar argument shows that $\Phi_0 R^d \in \mathcal{N}^e$ provided $R^d \in \mathcal{N}^d$. Note that the condition for the membership of R^e in $\mathcal{N}^e$ as given in [**10**], 1.1.11, that is (this time omitting "$\forall K \subset\subset \Omega \ \forall \alpha \in \mathbb{N}_0^s$")

$$\exists N \ \exists \gamma : \mathbb{N} \to \mathbb{R} \ \forall q \geq N \ \forall \varphi \in \mathcal{A}_q(\mathbb{R}^s) : \sup_{x \in K} |\partial^\alpha (R^e(S_\varepsilon \varphi, x))| = O(\varepsilon^{\gamma(q)-N})$$

(where $\gamma(q) \nearrow \infty$) is easily seen to be equivalent to

$$\forall n \ \exists q \qquad \forall \varphi \in \mathcal{A}_q(\mathbb{R}^s) : \sup_{x \in K} |\partial^\alpha (R^e(S_\varepsilon \varphi, x))| = O(\varepsilon^n)$$

which has the same structure as the condition in 7.3 for R^d to belong to $\mathcal{N}^d$:

$$\forall n \ \exists q \qquad \forall \phi \in \mathcal{C}_b^\infty(I \times \Omega, \mathcal{A}_q(\mathbb{R}^s)) : \sup_{x \in K} |\partial^\alpha (R^d(S_\varepsilon \phi(\varepsilon, x), x))| = O(\varepsilon^n) \ .$$

Due to the invariance of $\mathcal{E}_M$ and $\mathcal{N}$ under Φ_0, Φ_0 induces a map $\Phi : \mathcal{G}^d(\Omega) \to \mathcal{G}^e(\Omega)$ acting on representatives as restriction from $T^{-1}(\mathcal{A}_0(\Omega) \times \Omega)$ to $T^{-1}(\mathcal{A}_1(\Omega) \times \Omega)$. Φ is an algebra homomorphism respecting the embeddings of $\mathcal{D}'(\Omega)$ and differentiation.

REMARK 15.1. (i) If we had chosen to set $U^e(\Omega) = T^{-1}(\mathcal{A}_0(\Omega) \times \Omega)$ (contrary to [**10**], cf. the footnote to (15.1) above) j would be the identity map on $U^d(\Omega) = U^e(\Omega)$ and Φ_0 would be set-theoretic inclusion, hence injective.

(ii) Regarding the question of injectivity of Φ, the fact that $\mathcal{A}_1(\Omega)$ has been used in [**10**] and in (15.1) above to define $U^e(\Omega)$ (as compared to $\mathcal{A}_0(\Omega)$ in [**26**] for defining $U^d(\Omega)$) is completely irrelevant: Although this choice renders Φ_0 non-injective in general (consider $(0 \neq) R \in \mathcal{C}^\infty(U^d(\mathbb{R})) = \mathcal{C}^\infty(\mathcal{A}_0(\mathbb{R}) \times \mathbb{R})$ given by $(\varphi, x) \mapsto \int \xi \varphi(\xi) \, d\xi$: $\Phi_0 R = 0$ by the very definition of $U^e(\mathbb{R}) = \mathcal{A}_1(\mathbb{R}) \times \mathbb{R}$), $\mathcal{M} := \ker \Phi_0$ is contained in $\mathcal{N}^d$ since each $R \in \ker \Phi_0$ vanishes identically on pairs $(\phi(\varepsilon, x), x)$ where ϕ is a test object taking values in $\mathcal{A}_q(\mathbb{R})$ ($q \geq 1$). Thus the canonical image of $\mathcal{M}$ in $\mathcal{G}^d := \mathcal{E}_M^d / \mathcal{N}^d$ is trivial.

Having discussed Φ in detail, we will omit j and Φ_0 from our notation in the sequel. Now we can state precisely which properties a function $R : U^d(\Omega) \to \mathbb{C}$ has to satisfy if it is to refute the injectivity of Φ:

(i) $R \in \mathcal{E}^d$, i.e., R has to be smooth;
(ii) $R \in \mathcal{E}_M^d$,
(iii) $R \notin \mathcal{N}^d$,
(iv) $R \in \mathcal{N}^e$.

In the following, we will define maps $P, Q : U^d(\mathbb{R}) \to \mathbb{C}$ each of which satisfies (i)–(iv) above, thereby providing a counterexample to the conjecture of the canonical map Φ being injective. We will give the complete argument for P while only indicating how to adapt the proof to get the analogous result for Q.

For the definition of P, Q let $s := 1$, $\Omega := \mathbb{R}$. We continue using the C-formalism. Although now $U(\Omega) = \mathcal{A}_0(\mathbb{R}) \times \mathbb{R} = \mathcal{A}_0(\Omega) \times \Omega$ note that the C-formalism, nevertheless, differs from the J-formalism with respect to embedding

$\mathcal{D}'$, differentiation, testing (which involves T in the case of the J-formalism) and, finally, with respect to the action induced by a diffeomorphism. As a prerequisite for writing down P, Q explicitly, we introduce the following notation:

$$\langle \varphi | \varphi \rangle := \int \varphi(\xi) \overline{\varphi(\xi)} \, d\xi \qquad (\varphi \in \mathcal{D}(\mathbb{R}))$$

$$v_k \in \mathcal{D}'(\mathbb{R}): \quad \langle v_k, \varphi \rangle := \int \xi^k \varphi(\xi) \, d\xi \qquad (\varphi \in \mathcal{D}(\mathbb{R}),\ k \in \mathbb{N}_0)$$

$$v_{\frac{1}{2}} \in \mathcal{D}'(\mathbb{R}): \quad \langle v_{\frac{1}{2}}, \varphi \rangle := \int |\xi|^{\frac{1}{2}} \varphi(\xi) \, d\xi \qquad (\varphi \in \mathcal{D}(\mathbb{R}))$$

$$v(\varphi) := \langle \varphi | \varphi \rangle^{\frac{1}{2}} \langle v_{\frac{1}{2}}, \varphi \rangle \qquad (\varphi \in \mathcal{D}(\mathbb{R}))$$

$$g(x) := \frac{x}{1+x^2} \qquad (x \in \mathbb{R})$$

$$e(x) := \begin{cases} \exp(-\frac{1}{x}) & (x > 0) \\ 0 & (x \le 0) \end{cases} \qquad (x \in \mathbb{R})$$

$$\gamma_k := k + \frac{1}{k} \qquad (k \in \mathbb{N}).$$

Finally, choose an (even) function $\sigma \in \mathcal{D}(\mathbb{R})$ satisfying $0 \le \sigma \le 1$, $\sigma(x) \equiv 1$ for $|x| \le \frac{1}{2}$, $\sigma(x) \equiv 0$ for $|x| \ge \frac{3}{2}$ and set

$$h_k(x) := \sigma(x) \cdot 2g(x) + (1 - \sigma(x)) \cdot \operatorname{sgn}(x) \cdot |2g(x)|^{\gamma_k} \qquad (x \in \mathbb{R},\ k \in \mathbb{N}).$$

Being bounded and linear resp. bilinear (over $\mathbb{R}$), v_k, $v_{\frac{1}{2}}$ and $\langle . | . \rangle$ are smooth on $\mathcal{D}(\mathbb{R})$ ($k \in \mathbb{N}_0$). On $\mathcal{A}_0(\mathbb{R})$, $\langle \varphi | \varphi \rangle > 0$. Thus v is smooth on $\mathcal{A}_0(\mathbb{R})$ as a product of smooth functions. Observe that $\mathcal{A}_q(\mathbb{R}) = \mathcal{A}_0(\mathbb{R}) \cap \bigcap_{k=1}^{q} \ker v_k$.

In the sequel, we will make use of the following facts concerning g and e: For every $n \in \mathbb{N}_0$ there exists a constant $c_n > 0$ such that for all $x \ne 0$

$$|g^{(n)}(x)| \le \frac{c_n}{|x|^{n+1}}.$$

The derivatives of e can be written in the following form:

$$e^{(n)}(x) = e(x) \cdot \frac{q_n(x)}{x^{2r}} = \begin{cases} \exp(-\frac{1}{x}) \cdot \frac{q_n(x)}{x^{2r}} & (x > 0) \\ 0 & (x \le 0) \end{cases}$$

for every $n \in \mathbb{N}$ where q_n is a polynomial of degree $n - 1$ and $\frac{0}{0} := 0$.

Scaling of φ produces the following relations:

$$\langle S_\varepsilon \varphi | S_\varepsilon \varphi \rangle = \tfrac{1}{\varepsilon} \langle \varphi | \varphi \rangle$$

$$\langle v_k, S_\varepsilon \varphi \rangle = \varepsilon^k \langle v_k, \varphi \rangle$$

$$\langle v_{\frac{1}{2}}, S_\varepsilon \varphi \rangle = \varepsilon^{\frac{1}{2}} \langle v_{\frac{1}{2}}, \varphi \rangle$$

$$v(S_\varepsilon \varphi) = v(\varphi).$$

Apart from abbreviating $R \circ S^{(\varepsilon)} = R \circ (S_\varepsilon \times \operatorname{id})$ as R_ε for any function R defined on $\mathcal{A}_0(\mathbb{R}) \times \mathbb{R}$, we also will write R_ε for $R \circ S_\varepsilon$ if R is defined on $\mathcal{A}_0(\mathbb{R})$.

DEFINITION 15.2. Let $\varphi \in \mathcal{A}_0(\mathbb{R})$, $x \in \mathbb{R}$ and set

(15.2) $$P(\varphi, x) := \sum_{k=1}^{\infty} \frac{1}{k!} \cdot g\big(\langle \varphi | \varphi \rangle^{\gamma_k} e(v(\varphi))\big) \cdot \langle \varphi | \varphi \rangle^{\gamma_k} \cdot \langle v_k, \varphi \rangle,$$

(15.3) $$Q(\varphi, x) := \sum_{k=1}^{\infty} \frac{1}{k!} \cdot h_k\big(\langle \varphi | \varphi \rangle^{\frac{3}{2}} \langle v_{\frac{1}{2}}, \varphi \rangle\big) \cdot \langle \varphi | \varphi \rangle^{\gamma_k} \cdot \langle v_k, \varphi \rangle.$$

Hence P and Q, in fact, only depend on φ. We will see below that both series for P and Q converge uniformly on bounded subsets of $\mathcal{A}_0(\mathbb{R})$, making P and Q well-defined. For $k \in \mathbb{N}$, $\varphi \in \mathcal{A}_0(\mathbb{R})$ set

$$P_k(\varphi) := g\big(\langle\varphi|\varphi\rangle^{\gamma_k} e(v(\varphi))\big) \cdot \langle\varphi|\varphi\rangle^{\gamma_k} \cdot \langle v_k, \varphi\rangle.$$

Fix a positive number $\eta \leq 1$. To establish properties (i) and (ii) (i.e., smoothness and moderateness) of P, we will derive estimates of the form

(15.4) $$\quad |(\mathrm{d}^n(P_k)_\varepsilon)(\varphi)(\psi_1, \ldots, \psi_n)| \leq C_n^k \cdot \varepsilon^{-\frac{1}{k} - n\eta}$$

for some constants $C_n \geq 1$ not depending on ε ($n \in \mathbb{N}_0$), uniformly on any bounded subset B of $\mathcal{D}(\mathbb{R})$ and $\varphi \in B \cap \mathcal{A}_0(\mathbb{R})$, $\psi_1, \ldots, \psi_n \in B \cap \mathcal{A}_{00}(\mathbb{R})$. Setting $\varepsilon = 1$ in (15.4) shows that for each $n \in \mathbb{N}_0$, $(\mathrm{d}^n P_k)$ is an (eb)-sequence on $\mathcal{A}_0(\mathbb{R}) \times \mathcal{A}_{00}(\mathbb{R})^n$ which, by Theorem 14.4, implies smoothness of P. Considering arbitrary values of $\varepsilon \in I$, on the other hand, will lead to the proof of moderateness of P.

15.1. Proof of the estimates (15.4)

Fix $n \in \mathbb{N}_0$, $\varepsilon \in I$ and $0 < \eta \leq 1$. Set

$$\begin{aligned} P_k^{(1)}(\varphi) &:= g\big(\langle\varphi|\varphi\rangle^{\gamma_k} e(v(\varphi))\big) \\ P_k^{(2)}(\varphi) &:= \langle\varphi|\varphi\rangle^{\gamma_k} \\ P_k^{(3)}(\varphi) &:= \langle v_k, \varphi\rangle. \end{aligned}$$

(15.4) is equivalent to saying that the functions $\varepsilon^{\frac{1}{k} + n\eta} \cdot \mathrm{d}^n(P_k)_\varepsilon$ form an (eb)-sequence, with the respective constants in the estimate independent of ε. In order to prove this, by Leibniz' rule for the differential of a product and by Proposition 14.3 it suffices to show that each of the sequences (indexed by $k \in \mathbb{N}$) $\varepsilon^{n\eta} \mathrm{d}^m(P_k^{(1)})_\varepsilon$, $\varepsilon^{\gamma_k} \mathrm{d}^m(P_k^{(2)})_\varepsilon = \mathrm{d}^m(P_k^{(2)})$ and $\varepsilon^{-k} \mathrm{d}^m(P_k^{(3)})_\varepsilon = \mathrm{d}^m(P_k^{(3)})$ is (eb) for $m \leq n$, independently of ε. In the following, we are going to verify these claims step by step, starting with the elementary building blocks of the series defining P resp. Q.

Remark. For $w \in \mathcal{D}'(\Omega)$, $(\mathrm{d}w)(\varphi)(\psi) = \langle w, \psi\rangle$ and $\mathrm{d}^l w = 0$ for $l \geq 2$, due to the linearity of w. $\langle .|. \rangle$ being bilinear over $\mathbb{R}$, we obtain $(\mathrm{d}\langle .|. \rangle)(\varphi)(\psi) = \langle\psi|\varphi\rangle + \langle\varphi|\psi\rangle$, $(\mathrm{d}^2\langle .|. \rangle)(\varphi)(\psi_1, \psi_2) = \langle\psi_1|\psi_2\rangle + \langle\psi_2|\psi_1\rangle$ and $\mathrm{d}^l \langle .|. \rangle = 0$ for $l \geq 3$.

PROPOSITION 15.3. *The following sequences of functions of φ (indexed by $k \in \mathbb{N}$) are (eb) (1. and 3. on $\mathcal{D}(\mathbb{R})$, 2. on $\mathcal{A}_0(\mathbb{R})$):*
1. $\langle \xi^k, \varphi(\xi)\rangle$, $\langle |\xi|^k, \varphi(\xi)\rangle$, $\langle\varphi|\varphi\rangle^k$, $\langle\varphi|\varphi\rangle^{\gamma_k}$;
2. $\langle\varphi|\varphi\rangle^{-k}$, $\langle\varphi|\varphi\rangle^{-\gamma_k}$;
3. $(\beta_k)_n := \beta_k(\beta_k - 1)\ldots(\beta_k - n + 1)$ *(for fixed $n \in \mathbb{N}_0$; $(\beta_k)_0 := 1$)*

where the numbers $\beta_k \in \mathbb{R}$ occurring in 3. only have to satisfy an estimate of the form $|\beta_k| \leq pk$ for some fixed $p \in \mathbb{N}$.

PROOF. Fix a bounded subset B of $\mathcal{D}(\mathbb{R})$ containing at least one $\varphi \neq 0$. Then there exists a bounded set $L \subseteq \mathbb{R}$ containing the supports of all $\varphi \in B$. Let $\mathrm{m}(L) > 0$ denote the Lebesgue measure of L and set $C_1 := \max(1, \sup_{\xi \in L} |\xi|)$, $C_2 := \max(1, \mathrm{m}(L))$. Moreover, $C_3 := \max(1, \sup_{\varphi \in B} \|\varphi\|_\infty)$ is finite. Now let $\varphi \in B$.

1. We have

$$\max\big(|\langle\xi^k, \varphi(\xi)\rangle|, |\langle|\xi|^k, \varphi(\xi)\rangle|\big) \leq C_1^k C_2 C_3 \leq (C_1 C_2 C_3)^k,$$

$$\langle\varphi|\varphi\rangle^k \leq (C_2 C_3^2)^k,$$
$$\langle\varphi|\varphi\rangle^{\gamma_k} \leq (C_2 C_3^2)^{k+1} \leq (C_2^2 C_3^4)^k.$$

2. The Schwarz inequality yielding $1 = \langle 1, \varphi\rangle \leq (\int_L 1)^{\frac{1}{2}} \|\varphi\|_2 = (m(L)\langle\varphi|\varphi\rangle)^{\frac{1}{2}}$, we conclude $\langle\varphi|\varphi\rangle \geq C_2^{-1}$ and from this, in turn,

$$\langle\varphi|\varphi\rangle^{-k} \leq C_2^k,$$
$$\langle\varphi|\varphi\rangle^{-\gamma_k} \leq (C_2^2)^k.$$

3. The case $n = 0$ being trivial, note that there exists $C_0 > 1$ such that $k^n \leq C_0^k$ for all $k \in \mathbb{N}$. Consequently,

$$|(\beta_k)_n| \leq (|\beta_k| + n - 1)^n \leq (pk + n - 1)^n \leq (pkn)^n \leq (C_0 C)^k$$

where $C := \max(1, (pn)^n)$; the third inequality in the preceding chain is based on $0 \leq (pk - 1)(n - 1)$. □

Now the (ep)-property for $\varepsilon^{-k}\mathrm{d}^m(P_k^{(3)})_\varepsilon = \mathrm{d}^m(P_k^{(3)}) = \mathrm{d}^m(\langle v_k, \cdot\rangle)$ is clear from Proposition 15.3 and the remark preceding it. Regarding $\varepsilon^{\gamma_k}\mathrm{d}^m(P_k^{(2)})_\varepsilon = \mathrm{d}^m(P_k^{(2)}) = \mathrm{d}^m(\langle \cdot | \cdot \rangle^{\gamma_k})$ we obtain from the chain rule that $(\mathrm{d}^m\langle \cdot | \cdot \rangle^{\gamma_k})(\varphi)(\psi_1, \ldots, \psi_m)$ is given as a certain sum of terms of the form

$$(15.5) \qquad (\gamma_k)_l \langle\varphi|\varphi\rangle^{\gamma_k - l} \cdot (\mathrm{d}^{j_1}\langle \cdot | \cdot \rangle)(\ldots) \cdot \cdots \cdots \cdot \mathrm{d}^{j_l}\langle \cdot | \cdot \rangle)(\ldots),$$

the groups of three dots in parentheses standing for φ and certain subsequences of $(\psi_1, \ldots, \psi_m)$. ($j_1 + \cdots + j_l = m$ and, for any non-vanishing term of the above form, $j_1, \ldots, j_l \in \{1, 2\}$.) (15.5) immediately allows the application of Propositions 14.3 and 15.3, again in connection with the remark preceding the latter, thereby establishing the (eb)-property also for $\varepsilon^{\gamma_k}\mathrm{d}^m(P_k^{(2)})_\varepsilon$. For both terms treated so far, the constants occurring in the (eb)-estimate obviously are independent of ε. Observe that the case $n = 0$ of (15.4) is already settled completely on the basis of the results obtained so far, due to g being globally bounded on $\mathbb{R}$.

We now turn to the remaining one of the three terms which, to be sure, is the most difficult one to handle: We have to show $(\varepsilon^{n\eta}\mathrm{d}^m(P_k^{(1)})_\varepsilon)_k$ to constitute an (eb)-sequence, with the corresponding constant not depending on ε. Again according to the chain rule, the m-th differential of $\varepsilon^{n\eta}g(\varepsilon^{-\gamma_k}\langle\varphi|\varphi\rangle^{\gamma_k}e(v(\varphi)))$, evaluated at $\varphi; \psi_1, \ldots, \psi_m$, is given as a sum of terms of the form

$$\varepsilon^{n\eta - l\gamma_k}g^{(l)}(\varepsilon^{-\gamma_k}\langle\varphi|\varphi\rangle^{\gamma_k}e(v(\varphi))) \cdot f_1(\ldots) \cdot \cdots \cdots \cdot f_l(\ldots) \qquad (1 \leq l \leq m)$$

(maintaining the convention that a group of three dots at a differential's argument's place always denotes a certain subsequence of $(\varphi; \psi_1, \ldots, \psi_l)$) where each $f_{l'}(\ldots)$ is of the form

$$(15.6) \qquad \mathrm{d}^j(\langle \cdot | \cdot \rangle^{\gamma_k} \cdot (e \circ v))(\ldots) \qquad (1 \leq j \leq l \leq m).$$

On the basis of Leibniz' rule, Proposition 14.3 and the fact that $\langle\varphi|\varphi\rangle^{\gamma_k}$ together with all its differentials is already known to be (eb), it will suffice to deal with terms of the form

$$(15.7) \qquad \varepsilon^{n\eta - l\gamma_k}g^{(l)}(\varepsilon^{-\gamma_k}\langle\varphi|\varphi\rangle^{\gamma_k}e(v(\varphi))) \cdot \mathrm{d}^{i_1}(e \circ v)(\ldots) \cdot \cdots \cdots \cdot \mathrm{d}^{i_l}(e \circ v)(\ldots)$$

where $0 \leq i_{l'} \leq l$ and $i_1 + \ldots i_l \leq m$. To this end, we have to analyze $\mathrm{d}^i(e \circ v)$ for $0 \leq i \leq j \leq l \leq m$. Once more by the chain rule, this is a sum of products of $e^{(r)}(v(\varphi))$ with r factors which are differentials of v ($0 \leq r \leq i$). The proof of Proposition 15.3 shows that $\langle\varphi|\varphi\rangle^{\frac{1}{2} - r'}$ is bounded on bounded sets for $r' \in \mathbb{N}$.

15.1. PROOF OF THE ESTIMATES (15.4)

From this it follows that also the differentials of v are bounded on bounded sets; as they do not depend on k, they form an (eb)-sequence in a trivial manner. By Proposition 14.3 again we can discard them for the rest of the argument. Thus we are left with estimating

(15.8)
$$\varepsilon^{n\eta - l\gamma_k} g^{(l)}\left(\varepsilon^{-\gamma_k} \langle\varphi|\varphi\rangle^{\gamma_k} e(v(\varphi))\right) \cdot e^{(r_1)}(v(\varphi)) \cdot \ldots \cdot e^{(r_l)}(v(\varphi)) \qquad (1 \le l \le m)$$

where $0 \le r_t \le i_t \le l$ $(1 \le t \le l)$ and $r_1 + \cdots + r_l \le m$. Now $e^{(r)}(v(\varphi))$ can be written as

$$e^{(r)}(v(\varphi)) = e(v(\varphi)) \frac{q_r(v(\varphi))}{v(\varphi)^{2r}}$$

where q_r is a certain polynomial of degree $r - 1$. Consequently, (15.8) takes the form

$$\varepsilon^{n\eta - l\gamma_k} g^{(l)}(X) \cdot e(v(\varphi))^l \cdot \frac{1}{v(\varphi)^{2n}} \cdot v(\varphi)^{2(n-\bar{r})} \prod_{t=1}^{l} q_{r_t}(v(\varphi))$$

where we have set $X := \varepsilon^{-\gamma_k} \langle\varphi|\varphi\rangle^{\gamma_k} e(v(\varphi))$ and $\bar{r} := \sum_{t=1}^{l} r_t$, for the sake of brevity. Now expand $e(v(\varphi))^l$ according to

$$e(v(\varphi))^l = X^{l(1-\frac{\eta}{\gamma_k})} \cdot (\varepsilon^{\gamma_k} \langle\varphi|\varphi\rangle^{-\gamma_k})^{l(1-\frac{\eta}{\gamma_k})} \cdot e(v(\varphi))^{l\frac{\eta}{\gamma_k}}$$

and regroup the terms in the following way as to obtain the desired estimates:

1. Collecting all powers of ε, we obtain $\varepsilon^{n\eta - l\gamma_k} \cdot (\varepsilon^{\gamma_k})^{l(1-\frac{\eta}{\gamma_k})} = \varepsilon^{(n-l)\eta} \le 1$.

2. For $|X| \le 1$, we have $\left|g^{(l)}(X) \cdot X^{l(1-\frac{\eta}{\gamma_k})}\right| \le |g^{(l)}(X)| \le \|g^{(l)}\|_\infty$ (note that $0 < \eta \le 1 < \gamma_1 = 1 + 1 \le \gamma_k$ and that, consequently, $\frac{\eta}{\gamma_k} \le \frac{1}{2}$ for all $k \in \mathbb{N}$), while for $|X| \ge 1$ and c_l denoting a positive constant dominating $|x|^{l+1}|g^{(l)}(x)|$ for all $x \in \mathbb{R}$ (see the remarks after the introduction of g), we obtain

$$\left|g^{(l)}(X) \cdot X^{l(1-\frac{\eta}{\gamma_k})}\right| \le c_l |X|^{-l-1+l-l\frac{\eta}{\gamma_k}} \le c_l.$$

Altogether, the function $X \mapsto g^{(l)}(X) \cdot X^{l(1-\frac{\eta}{\gamma_k})}$ $(l = 1, \ldots, n)$ is globally bounded by a positive constant larger or equal to 1, say, C_g.

3. The following term, that is $\langle\varphi|\varphi\rangle^{-\gamma_k l(1-\frac{\eta}{\gamma_k})} = \langle\varphi|\varphi\rangle^{-l(\gamma_k - \eta)}$ gives rise to an (eb)-sequence letting $k = 1, 2, \ldots$: This is immediate from $\langle\varphi|\varphi\rangle^{-(\gamma_k - \eta)} = \langle\varphi|\varphi\rangle^{-\gamma_k} \cdot \langle\varphi|\varphi\rangle^{\eta}$ and Propositions 15.3 and 14.3, together with the observation that $\langle\varphi|\varphi\rangle^{\eta}$ is bounded on bounded sets. Hence for a given bounded subset B of $\mathcal{A}_0(\mathbb{R})$ there exists a constant $C_1 \ge 1$ satisfying $\langle\varphi|\varphi\rangle^{-l(\gamma_k - \eta)} \le C_1^k$ for all $\varphi \in B$, $k \in \mathbb{N}$.

4. $e(v(\varphi))^{l\frac{\eta}{\gamma_k}} \cdot \frac{1}{v(\varphi)^{2n}}$ can be rewritten as

$$e\left(\frac{\gamma_k v(\varphi)}{l\eta}\right) \cdot \left(\frac{l\eta}{\gamma_k v(\varphi)}\right)^{2n} \cdot \frac{1}{(l\eta)^{2n}} \cdot \gamma_k^{2n}.$$

Now $e(x) \cdot x^{-2n}$ (with $\frac{0}{0} := 0$) is globally bounded on $\mathbb{R}$ and $\gamma_k^{2n} \le (k+1)^{2n} \le (2k)^{2n} = 4^n k^{2n}$; the latter is (eb) by the proof of part 3 of Proposition 15.3. Therefore, $e(v(\varphi))^{l\frac{\eta}{\gamma_k}} \cdot \frac{1}{v(\varphi)^{2n}}$ can be estimated by C_2^k for a suitable constant $C_2 \ge 1$.

5. φ ranging over the bounded set B as in 3. above, $v(\varphi)$ attains values in a bounded subset of $\mathbb{C}$. On this set the polynomial $x^{2(n-\bar{r})} \prod_{t=1}^{l} q_{r_t}(x)$ is bounded by some constant $C_3 \geq 1$.

Summarizing, for any given bounded subset B of $\mathcal{A}_0(\mathbb{R})$ there exist constants C_g, C_1, C_2, C_3 (independent of $\varepsilon \in I$) such that

$$\left|\varepsilon^{n\eta - l\gamma_k} g^{(l)}\left(\varepsilon^{-\gamma_k} \langle \varphi | \varphi \rangle^{\gamma_k} e(v(\varphi))\right) \cdot e^{(r_1)}(v(\varphi)) \cdot \ldots \cdot e^{(r_l)}(v(\varphi))\right| \leq (C_g C_1 C_2 C_3)^k$$

for all $\varphi \in B$. This completes the proof of (15.4). □

15.2. Proof of smoothness of P

Setting $\varepsilon := 1$ in (15.4) shows $(\mathrm{d}^n P_k)$ to be an (eb)-sequence on $\mathcal{A}_0(\mathbb{R}) \times \mathcal{A}_{00}(\mathbb{R})^n$, for each $n \in \mathbb{N}_0$. Theorem 14.4 now implies that P as defined in 15.2 is smooth, that the differentials of P can be computed term-wise and that all the series for $\mathrm{d}^n P$ ($n \in \mathbb{N}_0$) converge with respect to τ_∞. □

15.3. Proof of moderateness of P

Let B be a bounded subset of $\mathcal{D}(\mathbb{R})$ and assume C_n ($n \in \mathbb{N}_0$) to be appropriate constants as to satisfy (15.4) for all $\varphi \in B \cap \mathcal{A}_0(\mathbb{R})$, $\psi_1, \ldots, \psi_n \in B \cap \mathcal{A}_{00}(\mathbb{R})$. Choosing $\eta := 1$, say, estimate (15.4) results in $|(\mathrm{d}^n(P_k)_\varepsilon)(\varphi)(\psi_1, \ldots, \psi_n)| \leq C_n^k \cdot \varepsilon^{-\frac{1}{k} - n}$. Multiplying by $\frac{1}{k!}$ and forming the infinite sum constituting $\mathrm{d}^n P_\varepsilon$, we obtain

$$|(\mathrm{d}^n P_\varepsilon)(\varphi)(\psi_1, \ldots, \psi_n)| \leq (e^{C_n} - 1) \cdot \varepsilon^{-1-n},$$

uniformly on B in the sense specified above. Hence P satisfies the condition equivalent to moderateness given in Theorem 7.12. □

We proceed to prove $P \notin \mathcal{N}^d$ resp. $P \in \mathcal{N}^e$. For the former negligibility property, instead of the condition as given in 7.3, we use the equivalent condition (once again omitting the quantifiers "$\forall K \subset\subset \Omega \ \forall \alpha \in \mathbb{N}_0^s$")

$$R \in \mathcal{N}^d \Leftrightarrow \forall n \ \exists q \ \forall B \text{ (bounded)} \subseteq \mathcal{A}_q(\mathbb{R}^s) : \sup_{x \in K, \ \varphi \in B} |\partial^\alpha(R(S_\varepsilon \varphi, x))| = O(\varepsilon^n)$$

occurring as $1°$ in Theorem 18 of [26]. Observe that for the application of the latter theorem, we need the fact that $P \in \mathcal{E}_M^d$ which has been shown above. For $\mathcal{N}^e$ we use the modified defining condition as given previously in this chapter (cf. the discussion of Φ_0):

$$R \in \mathcal{N}^e \Leftrightarrow \forall n \ \exists q \ \forall \varphi \in \mathcal{A}_q(\mathbb{R}^s) : \sup_{x \in K} |\partial^\alpha(R(S_\varepsilon \varphi, x))| = O(\varepsilon^n)$$

Clearly, our choice of the above forms of the respective conditions is motivated by the intention to have them as similar as possible to highlight the essential difference between them: The estimate on $|\partial^\alpha(R^e(S_\varepsilon \varphi, x))|$ is required to hold uniformly on bounded subsets with respect to φ in the former case as compared to only pointwise in the latter.

15.4. Proof of $P \notin \mathcal{N}^d$

Set $K := \{0\}$, $\alpha := 0$, $n := 1$. We are going to show that for this set of data the condition for P to belong to $\mathcal{N}^d$ is violated, i.e., we are going to show that for every $q \in \mathbb{N}$ there exists a bounded subset B of $\mathcal{A}_q(\mathbb{R})$ such that $\sup_{\varphi \in B} |(P(S_\varepsilon \varphi, 0))|$

is *not* of order $O(\varepsilon)$. To this end, let $q \in \mathbb{N}$. Since $v_{\frac{1}{2}}, v_0, v_1, \ldots, v_{q+1}$ are linearly independent in $\mathcal{D}'(\mathbb{R})$ there exist $\varphi_0, \varphi_1 \in \mathcal{A}_q(\mathbb{R})$ satisfying

$$\langle v_{\frac{1}{2}}, \varphi_0 \rangle = 0, \qquad \langle v_{q+1}, \varphi_0 \rangle = 1,$$
$$\langle v_{\frac{1}{2}}, \varphi_1 \rangle = 1, \qquad \langle v_{q+1}, \varphi_1 \rangle = 1.$$

Setting $\varphi_\lambda := (1-\lambda)\varphi_0 + \lambda \varphi_1$ ($0 \le \lambda \le 1$), $B := \{\varphi_\lambda \mid 0 \le \lambda \le 1\}$ is a bounded subset of $\mathcal{A}_q(\mathbb{R})$; moreover, $\langle v_{\frac{1}{2}}, \varphi_\lambda \rangle = \lambda$. For each λ in a suitable interval $(0, \lambda_0]$ we are going to specify some $\varepsilon_\lambda \in I$ with $\varepsilon_\lambda \to 0$ as $\lambda \to 0$ such that $P_{\varepsilon_\lambda}(\varphi_\lambda, 0) \to \infty$ ($\lambda \to 0$). Consequently, $\sup_{\varphi \in B} |(P(S_\varepsilon \varphi, 0))|$ is not even of order $O(1)$, i.e. not even bounded as $\varepsilon \to 0$. The (nonnegative) function defined by the assignment

$$\lambda \mapsto \varepsilon_\lambda := \langle \varphi_\lambda | \varphi_\lambda \rangle \cdot e(v(\varphi_\lambda))^{\frac{1}{\gamma_q+1}}$$

is continuous for $\lambda \in [0, 1]$, strictly positive for $\lambda > 0$ and satisfies $\varepsilon_0 = 0$. Hence there exists $\lambda_0 > 0$ such that $\varepsilon_\lambda \in I$ for $0 \le \lambda \le \lambda_0$. Moreover, $\varepsilon_\lambda \to 0$ as $\lambda \to 0$. The general term of the series defining $P_\varepsilon(\varphi_\lambda, 0)$ is given by (apart from the factor $\frac{1}{k!}$)

$$(P_k)_\varepsilon(\varphi_\lambda) = \varepsilon^{-\frac{1}{k}} \cdot g\big(\varepsilon^{-\gamma_k} \langle \varphi_\lambda | \varphi_\lambda \rangle^{\gamma_k} e(v(\varphi_\lambda))\big) \cdot \langle \varphi_\lambda | \varphi_\lambda \rangle^{\gamma_k} \cdot \langle v_k, \varphi_\lambda \rangle.$$

For $k = 1, \ldots, q$ this expression vanishes identically on B due to $\varphi_\lambda \in \mathcal{A}_q(\mathbb{R})$. For $k \ge q+2$ it can be estimated by $\varepsilon^{-\frac{1}{k}} \cdot \frac{1}{2} \cdot C^k$ (note that $\|g\|_\infty = \frac{1}{2}$) for some constant $C \ge 1$ being independent of λ since $\langle \varphi | \varphi \rangle^{\gamma_k} \langle v_k, \varphi \rangle$ forms an (eb)-sequence. Consequently,

$$\sum_{k=q+2}^\infty \frac{1}{k!} (P_k)_\varepsilon(\varphi_\lambda) \le \varepsilon^{-\frac{1}{q+2}} \cdot \frac{1}{2} \cdot e^C.$$

It remains to look at the leading term, that is, $(P_{q+1})_\varepsilon(\varphi_\lambda)$ (again omitting $\frac{1}{(q+1)!}$). Setting $\varepsilon := \varepsilon_\lambda$ it takes the value

$$\varepsilon_\lambda^{-\frac{1}{q+1}} \cdot g(1) \cdot \langle \varphi_\lambda | \varphi_\lambda \rangle^{\gamma_{q+1}} \cdot 1 \;=\; \varepsilon_\lambda^{-\frac{1}{q+1}} \cdot \frac{1}{2} \cdot \langle \varphi_\lambda | \varphi_\lambda \rangle^{\gamma_{q+1}}.$$

Altogether we obtain

$$\begin{aligned} P_{\varepsilon_\lambda}(\varphi_\lambda, 0) &\ge \frac{1}{(q+1)!} \cdot \varepsilon_\lambda^{-\frac{1}{q+1}} \cdot \frac{1}{2} \cdot \langle \varphi_\lambda | \varphi_\lambda \rangle^{\gamma_{q+1}} - \varepsilon^{-\frac{1}{q+2}} \cdot \frac{1}{2} \cdot e^C \\ &= \varepsilon_\lambda^{-\frac{1}{q+1}} \cdot \frac{1}{2} \left[\frac{\langle \varphi_\lambda | \varphi_\lambda \rangle^{\gamma_{q+1}}}{(q+1)!} - \varepsilon_\lambda^{\frac{1}{(q+1)(q+2)}} \cdot e^C \right] \end{aligned}$$

which tends to infinity as $\lambda \to 0$ (and, consequently, $\varepsilon_\lambda \to 0$), due to $\langle \varphi_\lambda | \varphi_\lambda \rangle$ being bounded from below uniformly for $\lambda \in [0, 1]$ and the second term in the square bracket vanishing in the limit. $\square$

15.5. Proof of $P \in \mathcal{N}^e$

Let $K \subset\subset \mathbb{R}$, $\alpha := 0$ (note that $P_\varepsilon(\varphi, x)$ does not depend on x) and $n \in \mathbb{N}$ be given; we claim that $q := n - 1$ is an appropriate choice for showing that

$$\forall \varphi \in \mathcal{A}_q(\mathbb{R}) : \sup_{x \in K} |P_\varepsilon(\varphi, x)| = O(\varepsilon^n).$$

Let $\varphi \in \mathcal{A}_q(\mathbb{R}) = \mathcal{A}_{n-1}(\mathbb{R})$. If $\langle v_{\frac{1}{2}}, \varphi \rangle \le 0$ then $v(\varphi) \le 0$ and, consequently, $e(v(\varphi)) = 0$ which in turn implies $P_\varepsilon(\varphi, x) = 0$ for all $x \in \mathbb{R}$ and all $\varepsilon \in I$. Thus we may assume that $\langle v_{\frac{1}{2}}, \varphi \rangle > 0$. But then also $v(\varphi)$ and, in turn, $e(v(\varphi))$ are positive.

Taking into account that $|g(x)| = |\frac{1}{x} \cdot \frac{x^2}{x^2+1}| \leq |\frac{1}{x}|$ for $x \neq 0$ we obtain the following estimate:

$$|g(\varepsilon^{-\gamma_k}\langle\varphi|\varphi\rangle^{\gamma_k} e(v(\varphi))) \cdot \varepsilon^{-\gamma_k}\langle\varphi|\varphi\rangle^{\gamma_k} \cdot \varepsilon^k \langle v_k,\varphi\rangle| \leq \varepsilon^k \frac{|\langle v_k,\varphi\rangle|}{e(v(\varphi))}.$$

Choosing a constant C satisfying $|\langle v_k,\varphi\rangle| \leq C^k$ for all $k \in \mathbb{N}$ (note that $(v_k)_k$ is (eb) by Proposition 15.3) we finally arrive at

$$|P_\varepsilon(\varphi,x)| \leq \sum_{k=q+1}^\infty \frac{1}{k!} \cdot \frac{\varepsilon^k C^k}{e(v(\varphi))} \leq \varepsilon^{q+1} \cdot \frac{e^C}{e(v(\varphi))}$$

thereby completing the proof of $P \in \mathcal{N}^e$. □

Now we turn to briefly discussing Q. In what follows we will tacitly assume all (eb)-questions to be handled appropriately. After scaling φ and dropping the factor $\frac{1}{k!}$ the typical term of the series defining Q takes the form

$$\varepsilon^{-\frac{1}{k}} \cdot h_k\left(\frac{1}{\varepsilon}\langle\varphi|\varphi\rangle\, v(\varphi)\right) \cdot \langle\varphi|\varphi\rangle^{\gamma_k} \cdot \langle v_k,\varphi\rangle.$$

As with P, $d^m(\langle\,.\,|\,.\,\rangle^{\gamma_k})$ and $d^m(\langle v_k,\,.\,\rangle)$ are (eb) for all $m \in \mathbb{N}_0$. Modulo some (eb)-arguments again, the non-trivial part of dealing with $d^m h_k\left(\frac{1}{\varepsilon}\langle\varphi|\varphi\rangle v(\varphi)\right)$ consists in getting to grips with $\varepsilon^{-l} h_k^{(l)}\left(\frac{1}{\varepsilon}\langle\varphi|\varphi\rangle v(\varphi)\right)$ for $l \leq m$. Thanks to the harmless leading factor ε^{-l} (as compared to $\varepsilon^{-l\gamma_k}$ in the case of P) it is sufficient to note that there exists some constant $C \geq 1$ satisfying $\|h_k^{(l)}\|_\infty \leq C^k$ for all $k \in \mathbb{N}$ and $0 \leq l \leq m$ (observe that σ and g are globally bounded together with all their derivatives). Summarizing, we obtain that for all $m \leq n$ the sequences (with respect to k) $\varepsilon^m d^m h_k\left(\frac{1}{\varepsilon}\langle\varphi|\varphi\rangle v(\varphi)\right)$ and, consequently,

$$\varepsilon^n \cdot d^n \left(h_k\left(\frac{1}{\varepsilon}\langle\varphi|\varphi\rangle v(\varphi)\right) \cdot \langle\varphi|\varphi\rangle^{\gamma_k} \cdot \langle v_k,\varphi\rangle\right)$$

are (eb), with the respective constants not depending on ε. From this, smoothness and moderateness of Q follow. To obtain the proof of $Q \notin \mathcal{N}^d$ from the proof of $P \notin \mathcal{N}^d$ simply replace the former definition of ε_λ by $\varepsilon_\lambda := \langle\varphi_\lambda|\varphi_\lambda\rangle^{\frac{3}{2}} \cdot \lambda$ and use the fact that $h_k(1) = \|h_k\|_\infty = 1$. Finally, to show that $Q \in \mathcal{N}^e$, fix $\varphi \in \mathcal{A}_q(\mathbb{R})$. The case $\langle v_{\frac{1}{2}},\varphi\rangle = 0$ being trivial, assume that $\langle v_{\frac{1}{2}},\varphi\rangle \neq 0$. For $\varepsilon \leq \frac{2}{3}\langle\varphi|\varphi\rangle^{\frac{3}{2}}|\langle v_{\frac{1}{2}},\varphi\rangle|$ we have

$$\left|h_k\left(\frac{1}{\varepsilon}\langle\varphi|\varphi\rangle^{\frac{3}{2}}\langle v_{\frac{1}{2}},\varphi\rangle\right)\right| = \left|2g\left(\frac{1}{\varepsilon}\langle\varphi|\varphi\rangle^{\frac{3}{2}}\langle v_{\frac{1}{2}},\varphi\rangle\right)\right|^{\gamma_k} \leq \varepsilon^{\gamma_k}\left(\frac{2}{\langle\varphi|\varphi\rangle^{\frac{3}{2}}|\langle v_{\frac{1}{2}},\varphi\rangle|}\right)^{\gamma_k}.$$

The rest of the argument is similar to that for P.

The reader might ask if it is indeed necessary to come up with counterexamples as complicated as P and Q certainly are. The author doubts that easier ones might be possible. This view is based on reflecting on the rôles each of the three factors constituting a single term of the series (for P, say) in fact has to play:

- $\langle v_k,\varphi\rangle$ distinguishes between the spaces $\mathcal{A}_q(\mathbb{R})$; this is crucial for the negligibility properties.
- $\langle\varphi|\varphi\rangle^{\gamma_k} = \langle\varphi|\varphi\rangle^k \cdot \langle\varphi|\varphi\rangle^{\frac{1}{k}}$, on the one hand, after scaling of φ compensates for the factor ε^k generated by scaling φ in $\langle v_k,\varphi\rangle$. On the other hand, it introduces a factor $\varepsilon^{-\frac{1}{k}}$ making the first non-vanishing term of the series the dominant one as $\varepsilon \to 0$.

- $g(\langle\varphi|\varphi\rangle^{\gamma_k} e(\langle v,\varphi\rangle))$ allows the pointwise vs. uniformly distinction being necessary to obtain $P \notin \mathcal{N}^d$, $P \in \mathcal{N}^e$. Though $g(\langle\varphi|\varphi\rangle^{\gamma_k}\langle v,\varphi\rangle)$ would suffice to achieve the latter, this alternative choice for the argument of g would produce, via the chain rule, a factor $\varepsilon^{-n(k+\frac{1}{k})}$ in the k-th term of $\mathrm{d}^n P_\varepsilon$ which would be disastrous for the moderateness of P. The function e (together with $\varepsilon^{-\gamma_k}$ in the argument of g) suppressing this unwanted factor, P becomes moderate in the end.

Similar arguments apply to Q.

CHAPTER 16

Classification of smooth Colombeau algebras between $\mathcal{G}^d(\Omega)$ and $\mathcal{G}^e(\Omega)$

16.1. The development leading from $\mathcal{G}^e(\Omega)$ to $\mathcal{G}^d(\Omega)$

This section, in fact, does justice to the title of this monograph by going back to the roots of Colombeau algebras constructed according to the scheme outlined in chapter 3. Surveying the range of algebras lying between the algebras $\mathcal{G}^d(\Omega)$ and (the smooth version of) $\mathcal{G}^e(\Omega)$, we will discuss, in particular, to which extent at least the definition of the algebra $\mathcal{G}^1(\Omega)$ of [13] (which can be located within that range) has to be modified to obtain diffeomorphism invariance. To be sure, the introduction of $\mathcal{G}^1(\Omega)$ has to be considered as the decisive step towards the construction of a diffeomorphism invariant Colombeau algebra. The result of our analysis will be the construction of a diffeomorphism invariant Colombeau algebra $\mathcal{G}^2(\Omega)$ which is closer to $\mathcal{G}^1(\Omega)$ than $\mathcal{G}^d(\Omega)$ is.

Apart from $\mathcal{G}^e(\Omega)$, all algebras to be considered in this and the subsequent chapter have $\mathcal{C}^\infty(U(\Omega))$ resp. $\mathcal{C}^\infty(\mathcal{A}_0(\Omega) \times \Omega)$ as their basic space. In particular, they are smooth algebras in the sense that representatives R have to be smooth also with respect to φ. The maps σ, ι, D_i and the actions induced by a diffeomorphism are defined as in 7.1 and 5.5–5.8, respectively. The algebras will differ, however, as to the type of test objects used for selecting the moderate resp. negligible members from the basic space. We begin by briefly reviewing the development leading from $\mathcal{G}^e(\Omega)$ via $\mathcal{G}^1(\Omega)$ to $\mathcal{G}^d(\Omega)$.

Features distinguishing $\mathcal{G}^1(\Omega)$ from $\mathcal{G}^e(\Omega)$:

(1.0) Smooth dependence of R on (φ, x) rather than arbitrary dependence on φ and smoothness only with respect to x.

(1.1) Dependence of test objects on ε.

(1.2) Asymptotically vanishing moments of test objects as compared to the stronger condition $\phi(\varepsilon) \in \mathcal{A}_q(\mathbb{R}^s)$ for all ε (which would be the naïve analog of $\varphi \in \mathcal{A}_q(\mathbb{R}^s)$ in the case of $\mathcal{G}^e(\Omega)$).

Features distinguishing $\mathcal{G}^d(\Omega)$ from $\mathcal{G}^1(\Omega)$:

(2.1) Dependence of test objects also on $x \in \Omega$ (in fact, smooth dependence).

(2.2) In testing for moderateness, test objects for $\mathcal{G}^d(\Omega)$ can take arbitrary values in $\mathcal{A}_0(\mathbb{R}^s)$, independently of any moment condition.

Let us analyze briefly how compelling the above changes in the definitions in fact are if a diffeomorphism invariant algebra is to be obtained. We refrain from questioning (1.0), i.e., smoothness of R with respect to (φ, x), as well as from questioning the smoothness of test objects with respect to x in the sense of (2.1). Both properties being used in the proof of diffeomorphism invariance in an essential way, they are absolutely necessary from a pragmatic point of of view to guarantee

16.1. THE DEVELOPMENT LEADING FROM $\mathcal{G}^e(\Omega)$ TO $\mathcal{G}^d(\Omega)$

the smoothness of $(\hat{\mu}R)(S_\varepsilon\tilde{\phi}(\varepsilon,\tilde{x}),\tilde{x}) = R(S_\varepsilon\phi(\varepsilon,\mu\tilde{x}),\mu\tilde{x})$ with respect to $\tilde{x}$ (see chapter 7), to be sure. Of course, this does not amount to say that we have a formal proof that for the diffeomorphism invariance of an algebra, smoothness of R with respect to φ or of test objects with respect to x are logically necessary.

Smoothness of test objects with respect to ε definitely is not an issue of striking importance: The equivalence of conditions (B) and (C) in Theorem 10.5 (resp. of conditions (B') and (C') in Theorem 10.6) shows that test objects of the form $\phi \in \mathcal{C}_b^{[\infty,\Omega]}(I \times \Omega, \mathcal{A}_0(\mathbb{R}^s))$ give rise to $\mathcal{G}^d(\Omega)$ (via using them for testing moderateness resp. negligibility) independently of the assumption of smoothness of $\varepsilon \mapsto \phi(\varepsilon,x)$. With the appropriate respective modifications of the proof, this statement is valid for all types of test objects being dependent on ε or (ε,x), that is, it is true for all nine types $[z,Y]$ where z is one of εx or ε (see below for the definition of these types).

Next, if for a given diffeomorphism $\mu : \tilde{\Omega} \to \Omega$ the induced map $\hat{\mu} : \mathcal{E}(\Omega) \to \mathcal{E}(\tilde{\Omega})$ is to extend the usual action μ^* induced by μ on distributions then we necessarily have to set $\hat{\mu} = \bar{\mu}^*$, i.e., $(\hat{\mu}R)(\tilde{\varphi},\tilde{x}) = R(\bar{\mu}(\tilde{\varphi},\tilde{x}))$ with $\bar{\mu}$ as defined in 5.7. For purposes of testing, we have, in turn, no other choice than to consider $\hat{\mu}_\varepsilon = \bar{\mu}_\varepsilon^*$ where

$$\bar{\mu}_\varepsilon(\tilde{\varphi},\tilde{x}) = \left(\tilde{\varphi}\left(\frac{\mu^{-1}(\varepsilon. + \mu\tilde{x}) - \tilde{x}}{\varepsilon}\right) \cdot |\det D\mu^{-1}(\varepsilon. + \mu\tilde{x})|, \mu\tilde{x}\right).$$

From $(\hat{\mu}_\varepsilon R)(\tilde{\varphi},\tilde{x}) = R(\bar{\mu}_\varepsilon(\tilde{\varphi},\tilde{x}))$ it is now evident that a moderate (resp. negligible) function R from the basic space has to accept test objects which are dependent on ε as well as on x ((1.1.) and (2.1)) if $\hat{\mu}R$ is still to be moderate (resp. negligible) (see the discussion preceding Theorem 7.14). (1.2) is compelling since the property that certain moments of a test object have to vanish simply is not invariant under $\bar{\mu}$ resp. $\bar{\mu}_\varepsilon$. The moments of the transformed test objects only vanish asymptotically. This has the further consequence that accepting (2.1) raises the question of how to handle asymptotically vanishing moments with respect to uniformity in $x \in \Omega$: Since all the definitions and theorems involve uniformity on compact subsets of Ω it seems reasonable to adopt this condition also for the asymptotically vanishing moment property, possibly even for all derivatives $\partial_x^\alpha \phi$ of a test object $\phi(\varepsilon,x)$. We will discuss several variants below.

So there only remains change (2.2) for which there seems to be no apparent necessity. To be sure, accepting (2.2) widens the range of permissible test objects, thereby in turn reducing $\mathcal{E}_M(\Omega)$ and $\mathcal{N}(\Omega)$ in size (see example 17.11 (i) below). Yet it has to be admitted that by this reduction, no generalized functions which are of interest either in the development of the theory or in applications are lost. Quite to the contrary, accepting (2.2) has the advantage that the definition of $\mathcal{E}_M(\Omega)$ becomes simpler and, above all, that considerable flexibility is gained in how to define $\mathcal{N}(\Omega)$, as respective glances at Theorems 7.9 and 13.1 reveal. Nevertheless, the preceding discussion leaves open the possibility that a diffeomorphism invariant Colombeau algebra $\mathcal{G}^2(\Omega)$ could be constructed avoiding (2.2). $\mathcal{G}^2(\Omega)$ would be closer to $\mathcal{G}^1(\Omega)$ than $\mathcal{G}^d(\Omega)$ is; the preceding considerations seem to suggest that passing from $\mathcal{G}^1(\Omega)$ to $\mathcal{G}^2(\Omega)$ would represent the minimal modification of $\mathcal{G}^1(\Omega)$ leading to a diffeomorphism invariant Colombeau algebra. In any case, a construction as envisaged above would yield a second example of a diffeomorphism invariant Colombeau algebra.

16.2. Classification of test objects

The term "test object" will always refer to some element of $\mathcal{C}_b^\infty(I \times \mathcal{A}_0(\mathbb{R}^s))$; apart from functions $\phi(\varepsilon, x)$, this formally also includes test objects of the form $\phi(\varepsilon)$ (depending only on ε) as well as elements φ of $\mathcal{A}_0(\mathbb{R}^s)$. From now on, we will write $\langle \xi^\alpha, \varphi(\xi) \rangle$ or even only $\langle \xi^\alpha, \varphi \rangle$ in place of $\int \xi^\alpha \varphi(\xi)\, d\xi$ for $\varphi \in \mathcal{D}(\mathbb{R}^s)$, $\alpha \in \mathbb{N}_0^s$.

DEFINITION 16.1. Let $q \in \mathbb{N}$. A function $\phi : I \to \mathcal{D}(\mathbb{R}^s)$ (possibly constant and/or depending also on other arguments, e.g., on $x \in \Omega$) is said to have vanishing moments of order q if $\langle \xi^\alpha, \phi(\varepsilon)(\xi) \rangle = 0$ for all $\alpha \in \mathbb{N}_0^s$ with $1 \leq |\alpha| \leq q$. It is said to have asymptotically vanishing moments of order q if $\langle \xi^\alpha, \phi(\varepsilon)(\xi) \rangle = O(\varepsilon^q)$ for all $\alpha \in \mathbb{N}_0^s$ with $1 \leq |\alpha| \leq q$. To which extent this estimate is assumed to hold uniformly with respect to, e.g., $x \in \Omega$ has to be specified separately (see below).

A function ϕ taking values in $\mathcal{A}_0(\mathbb{R}^s)$ has vanishing moments of order q if and only if it takes values in $\mathcal{A}_q(\mathbb{R}^s)$, actually. To obtain a classification of Colombeau algebras lying in the range between $\mathcal{G}^d(\Omega)$ and (the smooth version of) $\mathcal{G}^e(\Omega)$ we introduce the symbols defined in the following list. They are meant to refer to test objects or to respective notions of moderateness and negligibility (based on test objects of the corresponding type) or, finally, to Colombeau algebras defined as quotients of the respective spaces of moderate functions.

Parametrization of test objects:

[c]   test objects being single elements ("constants") of $\mathcal{A}_0(\mathbb{R}^s)$ resp. $\mathcal{A}_q(\mathbb{R}^s)$

[ε]   test objects depending only on $\varepsilon \in I$

[εx]   test objects depending on $\varepsilon \in I$ as well as on $x \in \Omega$

Moments of test objects:

[0]   test objects taking values in $\mathcal{A}_0(\mathbb{R}^s)$ without any restriction on moments

[A]   test objects having asymptotically vanishing moments (this symbol always has to refer to test objects of type [ε])

[V]   test objects having vanishing moments, i.e., taking values in some $\mathcal{A}_q(\mathbb{R}^s)$

The following symbols make sense only for test objects of type [εx]; each of them indicates asymptotically vanishing moments of test objects, possibly also of their derivatives $\partial_x^\alpha \phi(\varepsilon, x)$, with the following respective specifications:

[A_l]   uniformly on the particular $K \subset\subset \Omega$ on which R is being tested ("locally")

[A_g]   uniformly on each $L \subset\subset \Omega$ ("globally")

[A_l^∞]   all derivatives uniformly on the particular $K \subset\subset \Omega$ on which R is being tested

[A_g^∞]   all derivatives uniformly on each $L \subset\subset \Omega$

If the compact set K on which R is being tested and/or the order q of the (asymptotic) vanishing of moments is to be specified, K resp. q will be put as subscript(s) to the corresponding A-symbol, e.g., $[A_l]_{K,q}$. Parametrization symbols may be combined with (suitable) moment symbols. If in a composed symbol $[z, Y]$

Y is one of the A-symbols then $z = \varepsilon$ resp. $z = \varepsilon x$, being redundant, will be omitted frequently.

Obviously, $[A_g^\infty]_q$ implies $[A_l^\infty]_{K,q}$ (for any $K \subset\subset \Omega$) and $[A_g]_q$; each of the latter, in turn, implies $[A_l]_{K,q}$. As the examples below show, none of the reverse implications is true.

EXAMPLE 16.2. (i) Let $\Omega := \mathbb{R}$ and $K := [-1, +1]$. Define $\phi_1(\varepsilon, x) := \varphi + \varepsilon^q \sin(x|\ln\varepsilon|)\psi$ where $\varphi \in \mathcal{A}_q(\mathbb{R})$ and $\psi \in \mathcal{A}_{00}(\mathbb{R})$ with $\langle \xi^k, \psi(\xi) \rangle = \delta_{kq}$ for $k = 1, \ldots, q$. $\langle \xi^q, \partial_x \phi_1(\varepsilon, 0)(\xi) \rangle = \varepsilon^q |\ln \varepsilon|$ is not of order ε^q, yet every $\partial_x^n \phi_1(\varepsilon, x)$ has bounded image. Hence ϕ_1 is of type $[A_g]_q$ and of type $[A_l]_{K,q}$, yet neither of type $[A_g^\infty]_q$ nor of type $[A_l^\infty]_{K,q}$.

(ii) Let $K \subset\subset \Omega$ and set $\phi_2(\varepsilon, x) := \lambda(x)\varphi_1 + (1 - \lambda(x))\varphi_2$ where $\varphi_1 \in \mathcal{A}_q(\mathbb{R}^s)$, $\varphi_2 \in \mathcal{A}_0(\mathbb{R}^s) \setminus \mathcal{A}_1(\mathbb{R}^s)$ and $\lambda \in \mathcal{D}(\Omega)$ such that $0 \leq \lambda \leq 1$ and $\lambda \equiv 1$ on an open neighborhood of K. Then for $q \in \mathbb{N}$, ϕ_2 is both of types $[A_l]_{K,q}$ and $[A_l^\infty]_{K,q}$, yet neither of type $[A_g]_q$ nor of type $[A_g^\infty]_q$.

16.3. Classification of full smooth Colombeau algebras

In the following, we will use the symbols introduced above to classify smooth Colombeau algebras with respect to the type of test objects used for testing moderateness. From a combinatorial point of view, there are eleven ways of performing this test, each corresponding to one of the eleven types of test objects. The following diagram displays these variants and the relations between them. The arrows are to be read as implications between the corresponding notions of moderateness or as inclusion relations between the corresponding sets of moderate functions (and similarly, for negligibility, as far as types [A] and [V] are concerned). They are *not* representing implications between the properties of test objects being of the particular types; a diagram of the latter kind would have to have the arrows reversed, of course.

$$
\begin{array}{ccccc}
[\varepsilon x, 0] & \rightarrow & [\varepsilon, 0] & \rightarrow & [c, 0] \\
\downarrow & & & & \\
[\varepsilon x, A_l] & \rightarrow & [\varepsilon x, A_g] & & \downarrow \\
\downarrow & & \downarrow & & \downarrow \\
[\varepsilon x, A_l^\infty] & \rightarrow & [\varepsilon x, A_g^\infty] & \rightarrow & [\varepsilon, A] \\
& & \downarrow & & \downarrow \\
& & [\varepsilon x, V] & \rightarrow & [\varepsilon, V] & \rightarrow & [c, V]
\end{array}
$$

From the diagram, a useful extension of the characterizations of $R \in \mathcal{N}^d(\Omega)$ obtained so far[1] is immediate: Test objects in condition (4^∞) are of type $[A_l^\infty]_K$ where K is the compact set on which R is tested. This dependence on K of the class of admissible test objects (going back to [26], Theorem 18) might seem undesirable since this class rather ought to be defined universally for tests on arbitrary compact sets. However, it is clear from the diagram that our list of equivalent conditions could be extended by adding a further condition (5^∞), obtained from (4^∞) by

---
[1] (3°), (4^∞) in 7.9; (0°)–(2°) in 13.1; (A')–(Z') in 10.6; (C''), (Z'') in 10.7 (in each case assuming $R \in \mathcal{E}_M^d(\Omega)$, in addition).

replacing $[A_l^\infty]$ with $[A_g^\infty]$: It suffices to observe that $(3°)$ and (4^∞) are based on test objects of types $[\varepsilon x, V]$ and $[\varepsilon x, A_l^\infty]$, respectively. It will be a consequence of Corollary 16.8 below that a further extension by an analogous equivalent condition $(6°)$, referring to type $[A_g]$, can be achieved.

One more glance at the diagram allows to clearly identify the obstacle against the diffeomorphism invariance of the algebra $\mathcal{G}^1(\Omega)$ defined in [**13**]: The Lemma in section 3 of that article only shows the μ-transform of test objects of type $[\varepsilon, A]$ (being used in defining $\mathcal{G}^1(\Omega)$) to be of type $[\varepsilon x, A_g]$; yet, this is not sufficient for a positive outcome of the μ-transform of R being tested for, say, moderateness, provided R is assumed to be moderate. So it is Theorem (**T6**) of the blueprint outlined in chapter 3 which fails for $\mathcal{G}^1(\Omega)$.

Now, if $[X]$ and $[Y]$ are chosen from the set of the eleven types such that $[Y]$ is located "south to east" with respect to $[X]$ in the diagram above (i.e., if $\mathcal{E}_M[X] \subseteq \mathcal{E}_M[Y]$) and if, in addition, $[Y]$ is one of the types $[A]$ or $[V]$ then it easily checked that $\mathcal{E}_M[X]$ is an algebra ((**T2**)) containing $\mathcal{N}[Y] \cap \mathcal{E}_M[X]$ as an ideal ((**T3**)). Consequently, $\mathcal{E}_M[X]/(\mathcal{N}[Y] \cap \mathcal{E}_M[X])$ is an algebra. We shall refer to algebras arising in this way by the term "Colombeau-type algebras". Altogether there are 46 admissible choices of pairs $[X], [Y]$. In the following definition, we will specify eleven algebras of this kind, one for each type of moderateness. These will be the only ones we are to deal with in the sequel. They might be called "primary" since each of the remaining Colombeau-type algebras can be obtained as some subalgebra or some quotient algebra of one of them. Note, however, that the collection of these eleven algebras is not minimal in this respect (see Theorem 17.10).

DEFINITION 16.3. If $[X]$ is one of the types $[V]$ or $[A]$ define
$$\mathcal{G}[X] := \mathcal{E}_M[X]/\mathcal{N}[X];$$
for types $[0]$ define
$$\mathcal{G}[\varepsilon x, 0] := \mathcal{E}_M[\varepsilon x, 0]/(\mathcal{N}[\varepsilon x, A_l^\infty] \cap \mathcal{E}_M[\varepsilon x, 0]),$$
$$\mathcal{G}[\varepsilon, 0] := \mathcal{E}_M[\varepsilon, 0]/(\mathcal{N}[\varepsilon, A] \cap \mathcal{E}_M[\varepsilon, 0]),$$
$$\mathcal{G}[c, 0] := \mathcal{E}_M[c, 0]/(\mathcal{N}[c, V] \cap \mathcal{E}_M[c, 0]).$$
(The open set Ω has been omitted from the notation of the respective algebras.)

We will refer to $\mathcal{G}[X]$ also by "the algebra of type $[X]$". Each of the algebras mentioned at the beginning of this chapter is one of the eleven algebras just defined: Denoting by $\mathcal{G}_0^e(\Omega)$ the "smooth part" of $\mathcal{G}^e(\Omega)$, i.e., the subalgebra formed by all members having a smooth representative $R \in \mathcal{C}^\infty(U^e(\Omega))$, it is easy to see that $\mathcal{G}_0^e(\Omega) = \mathcal{G}[c, V]$. $\mathcal{G}^1(\Omega)$ obviously is equal to $\mathcal{G}[\varepsilon, A]$; the algebra $\mathcal{G}^2(\Omega)$ to be introduced in the following chapter is obtained as $\mathcal{G}[\varepsilon x, A_g^\infty]$. $\mathcal{G}^d(\Omega)$, finally, is given as $\mathcal{G}[\varepsilon x, 0]$. Observe that according to Theorem 7.9, $\mathcal{N}[\varepsilon x, A_l^\infty]$ can be replaced by $\mathcal{N}[\varepsilon x, V]$ in the definition of $\mathcal{G}[\varepsilon x, 0]$. Moreover, it should be clear from Example 7.7 and the discussion preceding them why $\mathcal{G}[\varepsilon x, 0]$ has *not* been defined as the quotient with respect to $\mathcal{N}[\varepsilon x, A_l] \cap \mathcal{E}_M[\varepsilon x, 0]$: This choice (corresponding to using condition $(4°)$ of [**26**], Theorem 18) would invalidate part (iii) of (**T1**) and thus prevent ι from preserving the product of smooth functions.

Corollary 16.8 below will show that test objects of types $[A_g]$ and $[A_g^\infty]$, respectively, give rise to the same moderate resp. negligible functions. Moreover, it will follow from Theorem 17.4 that also test objects of type $[A_l^\infty]$ lead to the same respective notions of moderateness and negligibility as test objects of type $[A_g^\infty]$ do. This actually leaves us with nine possibly different algebras.

The diagram formed by the canonical homomorphisms between the resulting nine algebras is not isomorphic to the previous diagram: On the one hand, as mentioned above, $[A_g]$, $[A_1^\infty]$ and $[A_g^\infty]$ have to be merged to represent $\mathcal{G}^2(\Omega) = \mathcal{G}[\varepsilon x, A_g^\infty]$. On the other hand, there is no canonical homomorphism from $\mathcal{G}^d(\Omega) = \mathcal{G}[\varepsilon x, 0]$ into $\mathcal{G}[\varepsilon x, A_1]$ since $\mathcal{N}[\varepsilon x, A_1] \cap \mathcal{E}_M[\varepsilon x, 0]$—not containing any of $R(\varphi, x) := \langle \xi^\beta, \varphi(\xi) \rangle$—is strictly smaller than $\mathcal{N}[\varepsilon x, A_1^\infty] \cap \mathcal{E}_M[\varepsilon x, 0]$. We do have canonical homomorphisms, however, both from $\mathcal{G}^d(\Omega) = \mathcal{G}[\varepsilon x, 0]$ and from $\mathcal{G}[\varepsilon x, A_1]$ into $\mathcal{G}^2(\Omega) = \mathcal{G}[\varepsilon x, A_g^\infty]$. So we finally arrive at

$$\begin{array}{ccccc}
\mathcal{G}[\varepsilon x, 0] & \to & \mathcal{G}[\varepsilon, 0] & \to & \mathcal{G}[c, 0] \\
\downarrow & & \downarrow & & \\
\mathcal{G}[\varepsilon x, A_1] \to \mathcal{G}[\varepsilon x, A_g^\infty] & \to & \mathcal{G}[\varepsilon, A] & & \downarrow \\
\downarrow & & \downarrow & & \\
\mathcal{G}[\varepsilon x, V] & \to & \mathcal{G}[\varepsilon, V] & \to & \mathcal{G}[c, V]
\end{array}$$

When establishing **(T1)**–**(T8)** for $\mathcal{G}^2(\Omega)$ in the following chapter we will survey briefly which of these theorems is true for each of the (seven) algebras apart from $\mathcal{G}^2(\Omega) = \mathcal{G}[\varepsilon x, A_g^\infty]$ and $\mathcal{G}^d(\Omega) = \mathcal{G}[\varepsilon x, 0]$. Let us anticipate at this point the facts concerning **(T7)** and **(T8)**, i.e., diffeomorphism invariance: The following counterexamples of moderate functions R for which $\hat{\mu}R$ fails to be moderate for some diffeomorphism μ definitely eliminate six of the nine algebras from the class of possibly diffeomorphism invariant ones, beyond any pragmatic reasoning regarding techniques of proof.

EXAMPLE 16.4. Let $\Omega := \mathbb{R}$.
(i) The example $R_0(\varphi, x) := \exp(i \exp(\langle \varphi | \varphi \rangle))$ presented in [26] shows that all three algebras of type $[\varepsilon, Y]$ (Y=0,A,V), as well as the one of type $[c, 0]$ are *not* diffeomorphism invariant.
(ii) Define $R_1(\varphi, x) := \langle \xi, \varphi(\xi) \rangle \cdot \exp(\langle \varphi | \varphi \rangle)$. Since R_1 vanishes on $\mathcal{A}_1(\mathbb{R}) \times \mathbb{R}$, it is moderate with respect to any type $[z, V]$. Under the action induced by the diffeomorphism $\mu(x) := x + e^x$ of $\mathbb{R}$ onto itself, R_1 is transformed to a function $\hat{\mu}R_1$ which is not moderate with respect to any type $[z, V]$ since the values attained by

$$(\hat{\mu}R_1)(S_\varepsilon \varphi, x) = \exp\left(\frac{1}{\varepsilon} \int \frac{|\varphi(\xi)|^2}{1 + e^x e^{\varepsilon \xi}} d\xi\right) \cdot \left[\varepsilon \int \xi \varphi(\xi) d\xi + e^x \int (e^{\varepsilon \xi} - 1) \varphi(\xi) d\xi\right]$$

are not of any order ε^{-N} ($n \in \mathbb{N}$) even for simple test objects of the form $\varphi \in \mathcal{A}_N(\mathbb{R})$. Therefore, $\hat{\mu}R_1$ does not pass the test for moderateness. This example excludes all types $[z, V]$ from the class of diffeomorphism invariant algebras.

The details are left to the reader.

Thus we are left with the algebras of types $[\varepsilon x, 0]$, $[\varepsilon x, A_g^\infty]$ (together with the two equivalent types mentioned above) and $[\varepsilon x, A_1]$ as possible candidates for being diffeomorphism invariant. $[\varepsilon x, 0]$ giving rise to the algebra $\mathcal{G}^d(\Omega)$ introduced in chapter 7, we will define $\mathcal{G}^2(\Omega)$ in the following chapter on the basis of type $[\varepsilon x, A_g^\infty]$

and prove it to be a diffeomorphism invariant Colombeau algebra by establishing the corresponding Theorems **(T1)**–**(T8)**.

For a discussion of $\mathcal{G}[A_l]$, finally, we refer to the following chapter. It is clear from Example 7.7 that this algebra cannot be counted among the class of Colombeau algebras due to its multiplication not reproducing the product of smooth functions; moreover, we have to leave it open if it is a differential algebra at all since we do not know if $\mathcal{N}[A_l]$ is invariant under differentiation. Nevertheless, the spaces of moderate resp. negligible functions obtained from type $[\varepsilon x, A_l]$ test objects turn out to be diffeomorphism invariant. Despite the obvious faults of $\mathcal{G}[A_l]$, we have included type $[A_l]$ in our scheme, mainly to allow for a thorough discussion of condition (4°) of [**26**], Theorem 18.

Summarizing, the results of this and the following chapter show that $\mathcal{G}^d(\Omega)$ and $\mathcal{G}^2(\Omega)$ are the only diffeomorphism invariant Colombeau algebras among the eleven (resp. nine) algebras defined in 16.3.

To complete this chapter, it remains to prove that the tests based on types $[A_g^\infty]$ and $[A_g]$ are in fact equivalent. As demonstrated by Example 16.2 (i), there are test objects of type $[A_g]_q$ failing to be of type $[A_g^\infty]_q$. Nevertheless, both these classes of test objects do give rise to the same moderate resp. negligible functions. This fact will emerge as an immediate corollary from the following theorem.

THEOREM 16.5. *Let $\phi \in \mathcal{C}_b^\infty(I \times \Omega, \mathcal{A}_0(\mathbb{R}^s))$ and let $2 \leq q \in \mathbb{N}$. If ϕ is of type $[A_g]_q$ then it also is of type $[A_g^\infty]_{q-1}$.*

For the proof we need two lemmas.

LEMMA 16.6. *Let $c : I \times \Omega \to \mathbb{R}$ have second partial derivatives $\partial_i^2 c$ for some $i \in \{1, \ldots, s\}$ ($\partial_i = \frac{\partial}{\partial x_i}$). Let $q > 0$, $0 \leq r < 1$ and assume that $K \subset\subset L \subset\subset \Omega$. If
$\sup_{x \in L} |c(\varepsilon, x)| = O(\varepsilon^q)$ and $\sup_{x \in L} |\partial_i^2 c(\varepsilon, x)| = O(\varepsilon^{rq})$ then $\sup_{x \in K} |\partial_i c(\varepsilon, x)| = O(\varepsilon^{\frac{1+r}{2}q})$.*

PROOF. We consider values of $\varepsilon \in I$ which are less than $\operatorname{dist}(K, \partial L)$; set $p := q\frac{1-r}{2}$. For $x \in K$, $x + \varepsilon^p e_i \in L$. Taylor's Theorem yields
$$c(\varepsilon, x + \varepsilon^p e_i) = c(\varepsilon, x) + \varepsilon^p \partial_i c(\varepsilon, x) + \varepsilon^{2p}\frac{1}{2}\partial_i^2 c(\varepsilon, x_\theta)$$
where $x_\theta = x + \theta \varepsilon^p e_i$ for some $\theta \in (0,1)$; note that also $x_\theta \in L$. Consequently,
$$\partial_i c(\varepsilon, x) = \varepsilon^{-p} \underbrace{(c(\varepsilon, x + \varepsilon^p e_i) - c(\varepsilon, x))}_{O(\varepsilon^q)} - \varepsilon^p \underbrace{\frac{1}{2}\partial_i^2 c(\varepsilon, x_\theta)}_{O(\varepsilon^{rq})} = O(\varepsilon^{\frac{1+r}{2}q}),$$
uniformly for $x \in K$. □

For the second lemma, we inductively define a sequence of numbers r_k by setting $r_1 := 0$, $r_{k+1} := \frac{(1+r_k)^2}{4}$ ($k \in \mathbb{N}$). Being strictly increasing and bounded by 1, this sequence is convergent, its limit being equal to 1.

LEMMA 16.7. *For every $k \in \mathbb{N}$ the following holds: Let $c : I \times \Omega \to \mathbb{R}$ be smooth with respect to the variable x_i ($x = (x_1, \ldots, x_s) \in \Omega$) for some $i \in \{1, \ldots, s\}$. Let $q > 0$ and $K \subset\subset L \subset\subset \Omega$. If $\sup_{x \in L} |c(\varepsilon, x)| = O(\varepsilon^q)$ and $\sup_{x \in L} |\partial_i^m c(\varepsilon, x)| = O(1)$ for all $m \in \mathbb{N}$ then $\sup_{x \in K} |\partial_i c(\varepsilon, x)| = O(\varepsilon^{\frac{1+r_k}{2}q})$.*

PROOF. Proceeding by induction, the case $k = 1$ is immediate from Lemma 16.6 by setting $r := r_1 = 0$. Assume the statement of the lemma to be true for

16.3. CLASSIFICATION OF FULL SMOOTH COLOMBEAU ALGEBRAS

a particular $k \in \mathbb{N}$. Let c, i, q, K, L be as specified. Choose K_1, K_2 as to satisfy $K \subset\subset K_1 \subset\subset K_2 \subset\subset L$. From $\sup_{x \in L} |c(\varepsilon, x)| = O(\varepsilon^q)$ and $\sup_{x \in L} |\partial_i^m c(\varepsilon, x)| = O(1)$ for all $m \in \mathbb{N}$ we deduce, by assumption, $\sup_{x \in K_2} |\partial_i c(\varepsilon, x)| = O(\varepsilon^{\frac{1+r_k}{2}q})$. Applying the statement of the lemma (for the particular value of k under consideration) once more, this time to the function $\partial_i c$, with $\frac{1+r_k}{2}q$ in place of q and for the pair K_1, K_2 of compact sets, we obtain $\sup_{x \in K_1} |\partial_i^2 c(\varepsilon, x)| = O(\varepsilon^{(\frac{1+r_k}{2})^2 q})$. In a last step, we apply Lemma 16.6 to conclude that $\sup_{x \in K} |\partial_i c(\varepsilon, x)| = O(\varepsilon^{\bar{r}q})$ where $\bar{r} = \frac{1}{2}(1 + \frac{(1+r_k)^2}{4}) = \frac{1+r_{k+1}}{2}$, thereby showing the statement of the lemma to be true also for $k+1$. □

Proof of Theorem 16.5. Let $\phi \in \mathcal{C}_b^\infty(I \times \Omega, \mathcal{A}_0(\mathbb{R}^s))$ be of type $[A_g]_q$ where $2 \leq q \in \mathbb{N}$. Denoting $\langle \xi^\alpha, \phi(\varepsilon, x)(\xi) \rangle$ by $c_\alpha(\varepsilon, x)$ ($\alpha \in \mathbb{N}_0^s$), we have to show that

$$\sup_{x \in K} |\langle \xi^\alpha, \partial_x^\beta \phi(\varepsilon, x)(\xi) \rangle| = \sup_{x \in K} |\partial^\beta c_\alpha(\varepsilon, x)| = O(\varepsilon^{q-1})$$

for $1 \leq |\alpha| \leq q-1$ and all $K \subset\subset \Omega$, $\beta \in \mathbb{N}_0^s$. Fix $\alpha \in \mathbb{N}_0^s$ satisfying $1 \leq |\alpha| \leq q$. By assumption, we have $\sup_{x \in L} |c_\alpha(\varepsilon, x)| = O(\varepsilon^q)$ and $\sup_{x \in L} |\partial^\beta c_\alpha(\varepsilon, x)| = O(1)$ for all $L \subset\subset \Omega$ and all $\beta \in \mathbb{N}_0^s$. Since $q\frac{1+r_k}{2} \to q$ as $k \to \infty$, Lemma 16.7 yields that $\sup_{x \in K} |\partial_i c_\alpha(\varepsilon, x)| = O(\varepsilon^{q-\frac{1}{2}})$ for every $K \subset\subset \Omega$ and any $i = 1, \ldots, s$. Noting that also $(q - \frac{1}{2})\frac{1+r_k}{2} \to (q - \frac{1}{2})$, the same argument, applied to $\partial_i c_\alpha$ and ∂_j ($j = 1, \ldots, s$) in place of c_α and ∂_i, respectively, shows that $\sup_{x \in K} |\partial_j \partial_i c_\alpha(\varepsilon, x)| = O(\varepsilon^{q-(\frac{1}{2}+\frac{1}{4})})$, again for every $K \subset\subset \Omega$ and any $i, j = 1, \ldots, s$. By induction, we obtain $\sup_{x \in K} |\partial^\beta c_\alpha(\varepsilon, x)| = O(\varepsilon^{q-q_\beta})$ for all $\beta \in \mathbb{N}_0^s$ where $q_\beta = \sum_{i=1}^{|\beta|} 2^{-i} < 1$. From this we finally conclude that $\sup_{x \in K} |\partial^\beta c_\alpha(\varepsilon, x)| = O(\varepsilon^{q-1})$ for all $\beta \in \mathbb{N}_0^s$ and all $K \subset\subset \Omega$. □

COROLLARY 16.8. *Let $R \in \mathcal{E}(\Omega)$. R is moderate (resp. negligible) with respect to type $[A_g]$ if and only if it is moderate (resp. negligible) with respect to type $[A_g^\infty]$.*

PROOF. Necessity of the condition being obvious, let us show sufficiency. Assuming R to be moderate with respect to type $[A_g^\infty]$, fix $\alpha \in \mathbb{N}_0^s$, $K \subset\subset \Omega$. Choose $N_1 \in \mathbb{N}$ such that $\partial^\alpha(R(S_\varepsilon \phi_1(\varepsilon, x), x)) = O(\varepsilon^{-N_1})$ holds for every test object ϕ_1 of type $[A_g^\infty]_{N_1}$, uniformly on K. Now set $N := N_1 + 1$ and pick a test object ϕ of type $[A_g]_N$. By Theorem 16.5, ϕ is of type $[A_g^\infty]_{N-1}$, i.e., of type $[A_g^\infty]_{N_1}$. Due to our choice of N_1, $\partial^\alpha(R(S_\varepsilon \phi(\varepsilon, x), x)) = O(\varepsilon^{-N_1})$ resp. $O(\varepsilon^{-N})$ follow. A similar argument applies to negligibility of R. □

CHAPTER 17

The algebra $\mathcal{G}^2$; classification results

The algebra $\mathcal{G}^2(\Omega)$ of type $[\varepsilon x, A_g^\infty]$ to be analyzed below results from the algebra $\mathcal{G}^1(\Omega) = \mathcal{G}[\varepsilon, A]$ of [13] by applying the minimal modification necessary to obtain diffeomorphism invariance. Recall that a test object $\phi \in \mathcal{C}_b^\infty(I \times \Omega, \mathcal{A}_0(\mathbb{R}^s))$ is said to be of type $[\varepsilon x, A_g^\infty]_q$ if $\sup_{x \in K} |\langle \xi^\alpha, \partial_x^\beta \phi(\varepsilon, x)(\xi) \rangle| = O(\varepsilon^q)$ for every $K \subset\subset \Omega$, $\beta \in \mathbb{N}_0^s$ and $\alpha \in \mathbb{N}_0^s$ with $1 \leq |\alpha| \leq q$. Moderateness resp. negligibility of $R \in \mathcal{E}(\Omega) = \mathcal{C}^\infty(U(\Omega))$ are defined as follows (where $K \subset\subset \Omega$ and $\alpha \in \mathbb{N}_0^s$):

DEFINITION 17.1. $R \in \mathcal{E}(\Omega)$ is moderate with respect to type $[\varepsilon x, A_g^\infty]$ if the following condition is satisfied:

$\forall K \ \forall \alpha \ \exists N \in \mathbb{N} \ \forall \phi \in \mathcal{C}_b^\infty(I \times \Omega, \mathcal{A}_0(\mathbb{R}^s))$ which are of type $[\varepsilon x, A_g^\infty]_N$:

$$\sup_{x \in K} |\partial^\alpha(R(S_\varepsilon \phi(\varepsilon, x), x))| = O(\varepsilon^{-N}).$$

DEFINITION 17.2. $R \in \mathcal{E}(\Omega)$ is negligible with respect to type $[\varepsilon x, A_g^\infty]$ if the following condition is satisfied:

$\forall K \ \forall \alpha \ \forall n \in \mathbb{N} \ \exists q \in \mathbb{N} \ \forall \phi \in \mathcal{C}_b^\infty(I \times \Omega, \mathcal{A}_0(\mathbb{R}^s))$ which are of type $[\varepsilon x, A_g^\infty]_q$:

$$\sup_{x \in K} |\partial^\alpha(R(S_\varepsilon \phi(\varepsilon, x), x))| = O(\varepsilon^n).$$

Since we are dealing with $\mathcal{G}^2(\Omega)$ exclusively in the following, we simply denote the sets of moderate resp. negligible functions in the sense of the preceding definitions by $\mathcal{E}_M(\Omega)$, $\mathcal{N}(\Omega)$. To establish $\mathcal{G}^2(\Omega)$ as a diffeomorphism invariant Colombeau algebra we have to convince ourselves that Theorems **(T1)**–**(T8)** of the scheme presented in chapter 3 are true on the basis of the preceding definitions (compare chapter 7 for the detailed elaboration of these theorems in the case of $\mathcal{G}^d(\Omega)$). Though our main interest will be focused on type $[A_g^\infty]$, of course, for each of **(T1)**–**(T8)** we will specify for which of the remaining types (apart from $[\varepsilon x, A_g^\infty]$ and $[\varepsilon x, 0]$) it holds as well.

To start with, (i) and (ii) of **(T1)** follow from the corresponding statements with respect to $\mathcal{G}^d(\Omega)$ (Theorem 7.4 (i),(ii)) for all types since $[\varepsilon x, 0]$ generates the smallest one of all spaces $\mathcal{E}_M[X]$. We already know from Example 7.7 that (iii) of **(T1)** is not satisfied for type $[A_l]$. For all the remaining types, however, the corresponding statement follows immediately from part (iii) of Theorem 7.4 by observing that $\mathcal{N}[\varepsilon x, V] \cap \mathcal{E}_M[\varepsilon x, 0] = \mathcal{N}[\varepsilon x, A_1^\infty] \cap \mathcal{E}_M[\varepsilon x, 0] = \mathcal{N}[\varepsilon x, A_g^\infty] \cap \mathcal{E}_M[\varepsilon x, 0]$ is contained in each space $\mathcal{N}[Y]$ where $[Y]$ is different from $[A_l]$. (The preceding equalities are due to Theorem 7.9 resp. to $(4^\infty) \Leftrightarrow (5^\infty)$ derived in the preceding chapter.) Finally, the proof of part (iv) of **(T1)** given in chapter 7 for $\mathcal{G}^d(\Omega)$ uses test objects of type $[c, V]$ (generating the largest one of all spaces $\mathcal{N}[X]$) and therefore is valid for all types.

Theorems **(T2)** and **(T3)** are immediate from Leibniz' rule for all types.

As it had been the case for $\mathcal{G}^d(\Omega)$, **(T4)**–**(T6)** are the hard ones to prove also for $\mathcal{G}^2(\Omega)$. Fortunately, **(T6)** can be taken from chapter 7 with only a slight modification, as we will see. For **(T4)** and **(T5)**, however, we need analogs of Theorems 7.12 and 7.13 for type $[A_g^\infty]$ allowing to express moderateness resp. negligibility of R in terms of differentials of R_ε. To this end, we have to introduce appropriate classes of sets corresponding to the bounded subsets $B \subseteq \mathcal{D}(\mathbb{R}^s)$ occurring in Theorems 7.12 and 7.13. For any closed affine subspace E_1 of a locally convex space E, let $\mathcal{C}_b^\infty(I, E_1)$ denote the set of all smooth maps $\varphi : I \to E_1$ having bounded image.

DEFINITION 17.3. Let $k \in \mathbb{N}_0$, $q \in \mathbb{N}$.
A (k,q)-class is a subset $\mathcal{B}$ of $\mathcal{C}_b^\infty(I, \mathcal{A}_0(\mathbb{R}^s)) \times \left[\mathcal{C}_b^\infty(I, \mathcal{A}_{00}(\mathbb{R}^s))\right]^k$ satisfying the following conditions:
  (i) The set $\{\psi_0(\varepsilon), \ldots, \psi_k(\varepsilon) \mid (\psi_0, \ldots, \psi_k) \in \mathcal{B},\, \varepsilon \in I\}$ is bounded in $\mathcal{D}(\mathbb{R}^s)$;
  (ii) $\sup_{(\psi_i)\in\mathcal{B}} \sup_{i=0,\ldots,s} |\langle \xi^\beta, \psi_i(\varepsilon)(\xi)\rangle| = O(\varepsilon^q)$ for all $\beta \in \mathbb{N}_0^s$ with $1 \leq |\beta| \leq q$.

Note that $(\psi_0, \ldots, \psi_k) \in \mathcal{C}_b^\infty(I, \mathcal{A}_0(\mathbb{R}^s)) \times \left[\mathcal{C}_b^\infty(I, \mathcal{A}_{00}(\mathbb{R}^s))\right]^k$ forms a (k,q)-class $\{(\psi_0, \ldots, \psi_k)\}$ (consisting of a single element) if and only if each of $\psi_0, \ldots, \psi_k$ has asymptotically vanishing moments of order q. The following results are established by combining techniques of the respective proofs of Theorem 17 of [**26**] and of Theorem 10.5

THEOREM 17.4. *Let $R \in \mathcal{E}(\Omega)$. R is moderate of type $[A_g^\infty]$ if and only if the following condition is satisfied:*

$$\forall K \subset\subset \Omega\ \forall \alpha \in \mathbb{N}_0^d\ \forall k \in \mathbb{N}_0\ \exists N \in \mathbb{N}\ \text{such that for each } (k,N)\text{-class } \mathcal{B}:$$

$$\sup_{(\psi_i)\in\mathcal{B}} \sup_{x \in K} |\partial^\alpha d_1^k R_\varepsilon(\psi_0(\varepsilon), x)(\psi_1(\varepsilon), \ldots, \psi_k(\varepsilon))| = O(\varepsilon^{-N}).$$

Moreover, for given K, α, k, N the preceding condition is satisfied for all (k,N)-classes $\mathcal{B}$ if and only if it is satisfied for all (k,N)-classes consisting of a single element $(\psi_0, \ldots, \psi_k)$. Therefore, the uniformity requirement with respect to $\mathcal{B}$ can as well be omitted from the characterization of moderateness given above.

THEOREM 17.5. *Let $R \in \mathcal{E}(\Omega)$. R is negligible of type $[A_g^\infty]$ if and only if the following condition is satisfied:*

$$\forall K \subset\subset \Omega\ \forall \alpha \in \mathbb{N}_0^d\ \forall k \in \mathbb{N}_0\ \forall n \in \mathbb{N}\ \exists q \in \mathbb{N}\ \text{such that for each } (k,q)\text{-class } \mathcal{B}:$$

$$\sup_{(\psi_i)\in\mathcal{B}} \sup_{x \in K} |\partial^\alpha d_1^k R_\varepsilon(\psi_0(\varepsilon), x)(\psi_1(\varepsilon), \ldots, \psi_k(\varepsilon))| = O(\varepsilon^n).$$

Moreover, for given K, α, k, n, q the preceding condition is satisfied for all (k,q)-classes $\mathcal{B}$ if and only if it is satisfied for all (k,q)-classes consisting of a single element $(\psi_0, \ldots, \psi_k)$. Therefore, the uniformity requirement with respect to $\mathcal{B}$ can as well be omitted from the characterization of negligibility given above.

The proofs of Theorems 17.4 and 17.5 are deferred to the end of this chapter.

COROLLARY 17.6. *Let $R \in \mathcal{E}(\Omega)$. R is moderate (resp. negligible) with respect to type $[A_g^\infty]$ if and only if it is moderate (resp. negligible) with respect to type $[A_1^\infty]$.*

PROOF. Sufficiency of the condition being obvious, let us show necessity. Supposing R to be moderate with respect to type $[A_g^\infty]$, the differentials of R satisfy the condition of Theorem 17.4. For testing R on some $K \subset\subset \Omega$ as to moderateness with respect to type $[\text{A}_1^\infty]$, we have to consider test objects just of that type. Now it is exactly the easy part of the very proof of 17.4 which shows that this test gives a positive answer. The same argument applies to negligibility. □

From the preceding Corollary and Corollary 16.8, we see that all three types $[\text{A}_g^\infty]$, $[\text{A}_g]$ and $[\text{A}_1^\infty]$ give rise to the same notions of moderateness resp. negligibility, hence to the same Colombeau algebras. This fact also constitutes one of the key ingredients for obtaining an intrinsic description of the algebra $\mathcal{G}^d(\Omega)$ on manifolds: The property of a test object living on the manifold to have asymptotically vanishing moments can be formulated in intrinsic terms, indeed ([**24**], Definition 3.5); yet it would be virtually unmanageable to deal with the latter property also for derivatives of this test object, which, of course, are to be understood in this general case as appropriate Lie derivatives with respect to smooth vector fields. Now Corollaries 16.8 and 17.6 allow to dispense with derivatives of test objects as regards the asymptotic vanishing of the moments, provided all $K \subset\subset \Omega$ are taken into account ([**24**], Corollary 4.5).

The next corollary might come as a bit of a surprise since we are already used to type $[\text{A}_1]$ displaying rather bad properties. Observe that it (necessarily, compare Example 7.7) only refers to moderateness. The case at hand seems to be the only one where a certain symmetry between $\mathcal{E}_M$ and $\mathcal{N}$ is broken.

COROLLARY 17.7. *Let $R \in \mathcal{E}(\Omega)$. R is moderate with respect to type $[\text{A}_1]$ if and only if it is moderate with respect to type $[\text{A}_g^\infty]$ (resp. $[\text{A}_1^\infty]$ resp. $[\text{A}_g]$).*

PROOF. Necessity of the condition being obvious this time, let us show sufficiency. Suppose R to be moderate with respect to type $[\text{A}_g^\infty]$ and let $K \subset\subset \Omega$, $\alpha \in \mathbb{N}_0^s$ be given. According to Theorem 17.4, choose $N \in \mathbb{N}$ such that for every $k = 0, 1, \ldots, |\alpha|$, for every $\beta \in \mathbb{N}_0^s$ with $0 \leq |\beta| \leq |\alpha|$ and for every (k,N)-class $\mathcal{B}$,

$$\sup_{(\psi_i) \in \mathcal{B}} \sup_{x \in K} |\partial^\alpha \mathrm{d}_1^k R_\varepsilon(\psi_0(\varepsilon), x)(\psi_1(\varepsilon), \ldots, \psi_k(\varepsilon))| = O(\varepsilon^{-N}).$$

For any test object ϕ of type $[\text{A}_1]_{K,N(1+|\alpha|)}$, it now follows

$$\sup_K |\partial^\alpha(R_\varepsilon(\phi(\varepsilon,x),x))|$$
$$= \sup_K \left| \sum_{\beta,m} (\partial^\beta \mathrm{d}_1^m R_\varepsilon)(\phi(\varepsilon,x),x)(\partial^{\gamma_1}\phi(\varepsilon,x), \ldots, \partial^{\gamma_m}\phi(\varepsilon,x)) \right|$$
$$= \sup_K \left| \sum_{\beta,m} (\partial^\beta \mathrm{d}_1^m R_\varepsilon)(\phi(\varepsilon,x),x)(\varepsilon^N \partial^{\gamma_1}\phi(\varepsilon,x), \ldots, \varepsilon^N \partial^{\gamma_m}\phi(\varepsilon,x)) \cdot \varepsilon^{-mN} \right|$$
$$= O(\varepsilon^{-N-|\alpha|N})$$

since, for every $m \in \mathbb{N}_0$, the finite sequences $(\phi(\varepsilon,x), \varepsilon^N \partial^{\gamma_1}\phi(\varepsilon,x), \ldots, \varepsilon^N \partial^{\gamma_m}\phi(\varepsilon, x))$ (with x ranging over K) form an (m,N)-class. □

Now the proofs of (**T4**) and (**T5**), that is, of the invariance of $\mathcal{E}_M[\text{A}_g^\infty]$ and $\mathcal{N}[\text{A}_g^\infty]$ with respect to differentiation, follow from Theorems 17.4 and 17.5 in precisely the same way as they have been achieved in chapter 7 for the building blocks of $\mathcal{G}^d(\Omega)$ by means of Theorems 7.12 and 7.13. Digressing once more from the proof of $\mathcal{G}^2(\Omega)$ being a diffeomorphism invariant Colombeau algebra, let us deal with invariance under differentiation for the remaining types of algebras: Types $[\text{A}_g]$ and $[\text{A}_1^\infty]$ as well as the case of $\mathcal{E}_M[\text{A}_1]$ are covered by Corollaries 16.8, 17.6 and 17.7, respectively. Moreover, it easy to check that (**T4**) and (**T5**) are true for all types $[\varepsilon]$ and $[c]$. An appropriate modification of Theorem 17 of [**26**] putting $\mathcal{A}_N(\mathbb{R}^s)$

resp. $\mathcal{A}_q(\mathbb{R}^s)$ in place of $\mathcal{A}_0(\mathbb{R}^s)$ and employing the techniques used in the proof of (A)$\Leftrightarrow$(C) of Theorem 10.5 establishes the respective results to hold also for type $[\varepsilon x, V]$. Type $[\varepsilon x, 0]$ being covered by chapter 7, we are left only with $\mathcal{N}[A_l]$ to be discussed. However, lacking an analog of Theorem 7.13 resp. of Theorem 17.5 for type $[A_l]$ we are not in a position to express negligibility of R with respect to this type in terms of differentials of R_ε. This tool, however, was the basis for deducing invariance under differentiation. So for the time being, we find $\mathcal{N}[A_l]$ to be the only one among all 11+8 (to be precise, 8+6 pairwise different) spaces $\mathcal{E}_M[X]$ resp. $\mathcal{N}[X]$ for which the invariance with respect to differentiation has to remain an open problem.

Finally, let us consider the question of diffeomorphism invariance. As an inspection of the structure of the proof of Theorem 7.14 reveals, this theorem actually shows the μ-transform of test objects of types $[A_l]$, $[A_g]$, $[A_l^\infty]$, $[A_g^\infty]$ to be of the same type again, respectively, thereby establishing **(T6)** in all four cases. Moreover, we see that on the basis of Corollaries 16.8 and 17.6 even a weaker version of the last statement of Theorem 7.14, referring only to types $[A_l]$ and $[A_g]$, would suffice to obtain diffeomorphism invariance for all four types $[\varepsilon x, A]$ (and, still, for $\mathcal{E}_M[\varepsilon x, 0]$; hence this would completely satisfy also the needs of chapter 7 dealing with $\mathcal{G}^d(\Omega)!$): Derivatives $\partial_x^\alpha \phi$ ($\alpha \neq 0$) could be dispensed with in Theorem 7.14 and its proof.

Recall that our proofs of **(T7)** and **(T8)** in chapter 7 (stating the invariance of moderateness resp. negligibility under the action induced by a diffeomorphism) were based on the equivalence of conditions (C) and (Z) in Theorem 10.5 (resp. of (C'') and (Z'') in Corollary 10.7) which, in turn, used the extension of paths $\phi(\varepsilon, x)$ provided by Proposition 10.4. Now each of the four types $[\varepsilon x, A]$ is preserved by the extension process $\phi \mapsto \tilde{\phi}$. Thus the respective analogs of (C)$\Leftrightarrow$(Z) can be shown to hold for all types $[\varepsilon x, A]$ by the methods employed in Theorem 10.5.

Now the proofs of **(T7)** and **(T8)**, respectively, are literally the same for all four types $[\varepsilon x, A]$ as for the algebra $\mathcal{G}^d(\Omega)$ treated in chapter 7. The proof of **(T8)** is even simpler in the present case since we do not have to bother with bridging the gap between vanishing moments (as used in Definition 7.3) and asymptotically vanishing moments (as occurring in Theorem 7.14) which has been accomplished in chapter 7 by means of the equivalence $(3°)\Leftrightarrow(4^\infty)$ provided by Theorem 7.9.

Summarizing, we have established

THEOREM 17.8. *$\mathcal{G}^2(\Omega)$ is a diffeomorphism invariant Colombeau algebra which can be obtained by using test objects of any of the types $[A_g^\infty]$, $[A_g]$ or $[A_l^\infty]$.*

Test objects of type $[A_l]$, on the other hand, give rise to a diffeomorphism invariant algebra which does not preserve the product of smooth functions via ι and for which it remains open if it is a differential algebra at all. Moreover, we have shown that each of the remaining six algebras (apart from $\mathcal{G}^d(\Omega) = \mathcal{G}[\varepsilon x, 0]$) satisfies **(T1)**–**(T5)**, yet fails to be diffeomorphism invariant.

Also for the algebra $\mathcal{G}^2(\Omega)$ it is true that in characterizing the negligibility of $R \in \mathcal{E}_M(\Omega)$ in terms of the differentials of R_ε, derivatives can be dispensed with, those with respect to φ as well as those with respect to x. The numbering of the conditions in the following theorem corresponds to that of Theorem 13.1.

THEOREM 17.9. *Let $R \in \mathcal{E}(\Omega)$ be moderate with respect to type $[A_g^\infty]$ (resp. with respect to types $[A_g]$, $[A_l^\infty]$). Then R is negligible with respect to any one of* *these types if and only if one of the following (equivalent) conditions is satisfied:*

(0_A°) $\forall K \subset\subset \Omega$ $\forall n \in \mathbb{N}$ $\exists q \in \mathbb{N}$ *such that for each $(0,q)$-class $\mathcal{B}$:*

$$\sup_{(\psi_0) \in \mathcal{B}} \sup_{x \in K} |R_\varepsilon(\psi_0(\varepsilon), x)| = O(\varepsilon^n).$$

(1_A°) $\forall K \subset\subset \Omega$ $\forall \alpha \in \mathbb{N}_0^d$ $\forall n \in \mathbb{N}$ $\exists q \in \mathbb{N}$ *such that for each $(0,q)$-class $\mathcal{B}$:*

$$\sup_{(\psi_0) \in \mathcal{B}} \sup_{x \in K} |\partial^\alpha R_\varepsilon(\psi_0(\varepsilon), x)| = O(\varepsilon^n).$$

(2_A°) $\forall K \subset\subset \Omega$ $\forall \alpha \in \mathbb{N}_0^d$ $\forall k \in \mathbb{N}_0$ $\forall n \in \mathbb{N}$ $\exists q \in \mathbb{N}$ *such that for each (k,q)-class $\mathcal{B}$:*

$$\sup_{(\psi_i) \in \mathcal{B}} \sup_{x \in K} |\partial^\alpha \mathrm{d}_1^k R_\varepsilon(\psi_0(\varepsilon), x)(\psi_1(\varepsilon), \ldots, \psi_k(\varepsilon))| = O(\varepsilon^n).$$

In each of the preceding conditions, the uniformity requirement with respect to $\mathcal{B}$ can as well be omitted without changing the content of the condition, regardless of the moderateness of R.

PROOF. Due to Corollaries 16.8 and 17.6 it does not matter which of the three types is being considered. For $R \in \mathcal{E}_M[A_g^\infty]$, (2_A°) is equivalent to negligibility with respect to $[A_g^\infty]$ by Theorem 17.5. $(2_A^\circ) \Rightarrow (1_A^\circ) \Rightarrow (0_A^\circ)$ is trivial; $(0_A^\circ) \Rightarrow (1_A^\circ)$ and $(1_A^\circ) \Rightarrow (2_A^\circ)$ can be established by carefully replacing bounded subsets of $\mathcal{A}_0(\mathbb{R}^s)$ resp. of $\mathcal{A}_{00}(\mathbb{R}^s)$ by appropriately chosen (k,q)-classes in the respective proofs of Theorem 13.1 and part $(1^\circ) \Rightarrow (2^\circ)$ of Theorem 18 of [26]. As far as the proof of $(1_A^\circ) \Rightarrow (2_A^\circ)$ (proceeding by induction with respect to k) is concerned, the most delicate task in this respect consists in choosing appropriate $(k+1,q)$- resp. $(k-1,q)$-classes $\mathcal{B}_{+1}, \mathcal{B}_{-1}$ to be used in connection with $\partial^\alpha \mathrm{d}_1^{k+1} R_\varepsilon$ resp. $\partial^\alpha \mathrm{d}_1^{k-1} R_\varepsilon$ when $\partial^\alpha \mathrm{d}_1^k R_\varepsilon$ is being evaluated on some (k,q)-class $\mathcal{B}$. To this end, define $(k+1,q)$- resp. $(k-1,q)$-classes $\mathcal{B}_{+1}, \mathcal{B}_{-1}$ by

$$\mathcal{B}_{+1} := \{(\psi_0 + t\psi_k, \psi_1, \ldots, \psi_k, \psi_k) \mid (\psi_0, \ldots, \psi_k) \in \mathcal{B},\ 0 \le t \le 1\},$$
$$\mathcal{B}_{-1} := \{(\psi_0 + t\psi_k, \psi_1, \ldots, \psi_{k-1}) \mid (\psi_0, \ldots, \psi_k) \in \mathcal{B},\ 0 \le t \le 1\}$$

to provide the appropriate arguments for $\partial^\alpha \mathrm{d}_1^{k+1} R_\varepsilon$ resp. $\partial^\alpha \mathrm{d}_1^{k-1} R_\varepsilon$. On the basis of this choice of $\mathcal{B}_{+1}, \mathcal{B}_{-1}$ the proof of Theorem 18 of [26] can be upgraded by introducing ε as additional parameter throughout as to establish $(1_A^\circ) \Rightarrow (2_A^\circ)$ of the theorem.

Note that in the proof of $(0_A^\circ) \Rightarrow (1_A^\circ)$ being obtained from the proof of Theorem 13.1 by introducing the parameter ε, Theorem **(T4)** which, in turn, is based on Theorem 17.4 has to be invoked to guarantee the moderateness of $\partial_i R$.

The last statement of the theorem follows from Theorem 17.5 since the corresponding statement thereof contains, among others, α and k as free variables. □

A glance at the preceding proof shows virtually all the substantial results of this article to be involved. Note that a $(0,q)$-class consisting of a single element ϕ is nothing else than (a singleton containing) a test object of type $[\varepsilon, A]_q$. The equivalence of $R \in \mathcal{N}[A_g^\infty]$ and condition (0_A°) without the uniformity clause (provided $R \in \mathcal{E}_M[A_g^\infty]$) now shows that for a function $R \in \mathcal{E}(\Omega)$ which is moderate with respect to type $[\varepsilon x, A_g^\infty]$, it amounts to the same to be negligible with respect to either type $[\varepsilon x, A_g^\infty]$ or $[\varepsilon, A]$. We will make use of this fact below. For the following theorem, recall that $\mathcal{G}_0^e$ denotes the smooth part of $\mathcal{G}^e$ (cf. the discussion following Definition 16.3).

17. THE ALGEBRA $\mathcal{G}^2$; CLASSIFICATION RESULTS

THEOREM 17.10. *Of the canonical maps $\mathcal{G}^d(\Omega) \to \mathcal{G}^2(\Omega) \to \mathcal{G}^1(\Omega) \to \mathcal{G}_0^e(\Omega)$ the first and the second one are injective whereas the third one is not. The four corresponding spaces of representatives (i.e., of moderate functions) are pairwise different.*

PROOF. The injectivity of the map $\mathcal{G}^d(\Omega) \to \mathcal{G}^2(\Omega)$ is equivalent to $\mathcal{N}[\varepsilon x, V] \cap \mathcal{E}_M[\varepsilon x, 0] = \mathcal{N}[\varepsilon x, A_g^\infty] \cap \mathcal{E}_M[\varepsilon x, 0]$. This, however, is accomplished by the extension $(3°) \Leftrightarrow (5^\infty)$ of Theorem 7.9 derived from the first diagram in chapter 16. The injectivity of $\mathcal{G}^2(\Omega) \to \mathcal{G}^1(\Omega)$, on the other hand, is equivalent to $\mathcal{E}_M[\varepsilon x, A_g^\infty] \cap \mathcal{N}[\varepsilon, A] = \mathcal{N}[\varepsilon x, A_g^\infty]$ which has been deduced previously from Theorem 17.9. Finally, to establish the non-injectivity of $\mathcal{G}^1(\Omega) \to \mathcal{G}_0^e(\Omega)$, we use the fact that the function P introduced in chapter 15 can be shown not to be negligible with respect to type $[\varepsilon, A]$, by techniques similar to those employed in chapter 15. The difference of the respective spaces of moderate functions should be clear from the following examples. □

EXAMPLE 17.11. Let $\Omega := \mathbb{R}$.
(i) Both $R_2(\varphi, x) := \exp(\langle \varphi|\varphi\rangle^2 \langle \xi, \varphi(\xi)\rangle)$ and $R_3(\varphi, x) := \exp(\varphi(0)^2 \langle \xi, \varphi\rangle)$ are moderate of type $[\varepsilon x, A_g^\infty]$, yet not of type $[\varepsilon x, 0]$.
(ii) The counterexample $R_0(\varphi, x) := \exp(i \exp(\langle \varphi|\varphi\rangle))$ (see [**26**]) which has already been mentioned in connection with the failure of algebras of type $[\varepsilon]$ to be diffeomorphism invariant (see 16.4(i)) has the property of being moderate of type $[\varepsilon, A]$ yet not of type $[\varepsilon x, A_g^\infty]$. The same holds true for R_5 to be defined below.
(iii) $R_4(\varphi, x) := \langle \xi, \varphi(\xi)\rangle \cdot \exp(\varphi(0))$ is moderate of type $[c, V]$, yet not of type $[\varepsilon, A]$. This also holds true for R_1 introduced as Example 16.4 (ii).
(iv) $R_5(\varphi, x) := \exp(-\langle \varphi|\varphi\rangle) \cdot \exp(i \exp(2\langle \varphi|\varphi\rangle))$ is of particular interest: It is not moderate with respect to any of the types $[\varepsilon x]$, however, it is even negligible with respect to all types $[\varepsilon]$ and $[c]$.

Again the proofs of the preceding claims are left to the reader.

It is a remarkable fact that answering the apparently harmless question of injectivity of the canonical maps in the last analysis involves quite a number of hard theorems: the extension $(3°) \Leftrightarrow (5^\infty)$ of Theorem 7.9 derived in chapter 16; Theorem 17.9 which, in turn, is based on part $(1°) \Leftrightarrow (2°)$ Theorem 18 of [**26**] and on Theorems 13.1, 17.4 and 17.5; finally, also the counterexample P of chapter 15 is among the ingredients of the argument. It remains to prove Theorems 17.4 and 17.5.

Proof of Theorem 17.4. To show sufficiency of the condition, suppose that the differentials of R_ε (where $R \in \mathcal{E}(\Omega)$) satisfy the property specified in the theorem. Consider a test object $\phi(\varepsilon, x)$ of type $[A_g^\infty]_N$ and set $\Phi(\varepsilon, x) := (\phi(\varepsilon, x), x)$. Expanding $\partial^\alpha(R_\varepsilon \circ \Phi)$ according to the chain rule shows that R is moderate of type $[A_g^\infty]$: It suffices to observe that the family of all finite sequences

$$(\phi(\varepsilon, y), \partial_y^{\beta_1} \phi(\varepsilon, y), \ldots, \partial_y^{\beta_l} \phi(\varepsilon, y))$$

forms an (l, N)-class if ε is considered as variable and y as a parameter taking values in some compact subset of Ω.

Conversely, for a function $R \in \mathcal{E}(\Omega)$ which is moderate with respect to type $[A_g^\infty]$, we will show that the assumption of R to violate the condition in the theorem leads to a contradiction. Thus suppose that there exist $K \subset\subset \Omega$, $\alpha \in \mathbb{N}_0^s$, $k \in \mathbb{N}_0^s$ such that for all $N \in \mathbb{N}$ there exists a (k,N)-class $\mathcal{B}$ such that

$$\sup_{K,\mathcal{B}} |\partial^\alpha \mathrm{d}_1^k R_\varepsilon(\psi_0(\varepsilon), x)(\psi_1(\varepsilon), \ldots, \psi_k(\varepsilon))| \tag{17.1}$$

is *not* of order ε^{-N}. By moderateness of R, there exists $N \in \mathbb{N}$ such that

$$\sup_K |\partial^{\alpha'}(R_\varepsilon(\phi(\varepsilon, x), x))| = O(\varepsilon^{-N}) \tag{17.2}$$

for all test objects ϕ of type $[\mathrm{A}_\mathrm{g}^\infty]_N$, where $\alpha' := \alpha + pe_s$, $p := \sum_{i=1}^{k}(|\alpha| + k^2 + i)$. Due to our hypothesis, there exists a (k,N)-class $\mathcal{B}$ such that (17.1) is not of order ε^{-N}. Having fixed $K, \alpha, k, N, \mathcal{B}$, we inductively define sequences $x^{(j)} \in K$, $(\psi_0^{(j)}, \ldots, \psi_k^{(j)}) \in \mathcal{B}$, $0 < \varepsilon_j < \frac{1}{j}$ (with $\varepsilon_{j+1} < \varepsilon_j$) ($j = 1, 2, \ldots$) such that the following inequalities hold for $j = 1, 2, \ldots$:

$$|\partial^\alpha \mathrm{d}_1^k R_{\varepsilon_j}(\psi_0^{(j)}(\varepsilon_j), x^{(j)})(\psi_1^{(j)}(\varepsilon_j), \ldots, \psi_k^{(j)}(\varepsilon_j))| \geq j \cdot \varepsilon_j^{-N} \tag{17.3}$$

(the technical details are similar to those in the proof of part (C)$\Rightarrow$(A) of Theorem 10.5). Let $(\lambda_j)_{j \in \mathbb{N}}$ be a partition of unity on I as in Lemma 10.1; for $(t_1, \ldots, t_k) \in \{0, 1, \ldots, k\}^k$, define

$$\phi_{t_1, \ldots, t_k}(\varepsilon, x) := \sum_{j=1}^{\infty} \lambda_j(\varepsilon) \cdot \left[\psi_0^{(j)}(\varepsilon) + \sum_{i=1}^{k} t_i \frac{(x_s - x_s^{(j)})^{|\alpha| + k^2 + i}}{(|\alpha| + k^2 + i)!} \cdot \psi_i^{(j)}(\varepsilon) \right].$$

Since $\sum_{i=1}^{k} t_i \frac{(x_s - x_s^{(j)})^{|\alpha| + k^2 + i}}{(|\alpha| + k^2 + i)!}$ is a polynomial in x and all $(\psi_0^{(j)}, \ldots, \psi_k^{(j)})$ are members of one particular (k,N)-class $\mathcal{B}$, $\phi_{t_1, \ldots, t_k}$ is a member of $\mathcal{C}_b^\infty(I, \mathcal{A}_0(\mathbb{R}^s))$ and, in addition, is of type $[\mathrm{A}_\mathrm{g}^\infty]_N$. From (17.2) we conclude that

$$\sup_K |\partial^{\alpha'}(R_\varepsilon(\phi_{t_1, \ldots, t_k}(\varepsilon, x), x))| = O(\varepsilon^{-N}). \tag{17.4}$$

Now we follow the combinatorial reasoning of the proof of Theorem 17 of [**26**] to derive the desired contradiction: Choosing numbers $c_0, \ldots, c_k$ satisfying the set of equations $\sum_{i=0}^{k} c_i \cdot i^m = \delta_{1m}$ ($m = 0, 1, \ldots, k$), let us form

$$\sum_{t_1=0}^{k} \cdots \sum_{t_k=0}^{k} c_{t_1} \ldots c_{t_k} \partial^{\alpha'}(R_\varepsilon(\phi_{t_1, \ldots, t_k}(\varepsilon, x), x)). \tag{17.5}$$

By (17.4), this expression is of order ε^{-N}, uniformly for $x \in K$. On the other hand, evaluating (17.5) at $\varepsilon := \varepsilon_j$, $x := x^{(j)}$ according to the chain rule results in a positive integer multiple of $\partial^\alpha \mathrm{d}_1^k R_{\varepsilon_j}(\psi_0^{(j)}(\varepsilon_j), x^{(j)})(\psi_1^{(j)}(\varepsilon_j), \ldots, \psi_k^{(j)}(\varepsilon_j))$, due to the delicate combinatorial argument of the proof of Theorem 17 of [**26**]. (17.5) being of order ε^{-N}, we conclude that

$$|\partial^\alpha \mathrm{d}_1^k R_{\varepsilon_j}(\psi_0^{(j)}(\varepsilon_j), x^{(j)})(\psi_1^{(j)}(\varepsilon_j), \ldots, \psi_k^{(j)}(\varepsilon_j))| \leq C \varepsilon_j^{-N} \qquad (j \geq j_0)$$

for some positive constant $C > 0$ and some $j_0 \in \mathbb{N}$. This, however, contradicts our choice of $x^{(j)}$, $\psi_i^{(j)}$. So the condition in the theorem, in fact, is necessary for R being moderate. (In a trivial way, the preceding reasoning also applies in the case $k = 0$ if all sums $\sum_{i=1}^{k}$ are set equal to 0.)

17. THE ALGEBRA $\mathcal{G}^2$; CLASSIFICATION RESULTS

To prove the last statement of the theorem, let K, α, k, N be given. Suppose, again by way of contradiction, the condition on the differentials of R_ε given in the theorem to be satisfied for (k, N)-classes consisting of a single element, yet to be violated for arbitrary (k, N)-classes, either with respect to the particular K, α, k, N at hand. Similarly to the reasoning of the main part of the proof, deduce from these hypotheses the existence of a (k, N)-class $\mathcal{B}$ and of sequences $0 < \varepsilon_{j+1} < \varepsilon_j < \frac{1}{j}$, $x^{(j)} \in K$, $(\psi_0^{(j)}, \ldots, \psi_k^{(j)}) \in \mathcal{B}$ $(j = 1, 2, \ldots)$ satisfying the inequalities (17.3) for all $j \in \mathbb{N}$. Now define

$$\psi_i(\varepsilon) := \sum_{j=1}^\infty \lambda_j(\varepsilon) \psi_i^{(j)}(\varepsilon) \qquad (i = 0, 1, \ldots, k)$$

where $(\lambda_j)_{j \in \mathbb{N}}$ is a partition of unity on I as in Lemma 10.1. Due to the properties of the $\psi_i^{(j)}$, $\{(\psi_0, \ldots, \psi_k)\}$ is a (k, N)-class. By assumption,

$$\sup_K |\partial^\alpha \mathrm{d}_1^k R_\varepsilon(\psi_0(\varepsilon), x)(\psi_1(\varepsilon), \ldots, \psi_k(\varepsilon))| = O(\varepsilon^{-N}).$$

Taking into account that $\psi_i(\varepsilon_j) = \psi_i^{(j)}(\varepsilon_j)$, this contradicts our choice of $x^{(j)}$, $\psi_i^{(j)}$, thereby completing the proof. □

Proof of Theorem 17.5. Just copy the proof of Theorem 17.4, add "for all $n \in \mathbb{N}$" at the appropriate places and change "ε^{-N}" to "ε^n". At the remaining occurrences of N, replace it by q. □

CHAPTER 18

Concluding remarks

As has been pointed out already in Part 1, Theorem 7.14 has a place at the very core of diffeomorphism invariance of a Colombeau algebra. The problem with algebras of any type $[c, Y]$, of course, is that classes consisting of test objects as simple as $\varphi \in \mathcal{A}_0(\mathbb{R}^s)$ are not invariant under the action of a diffeomorphism since the latter introduces dependence on ε and x. Types $[\varepsilon x, Y]$ of course are an efficient remedy against that problem as they incorporate a very general (ε, x)-dependence into test objects. Yet there is an intermediate way: Starting with the class of "constant" test objects $\tilde\varphi \in \mathcal{A}_0(\mathbb{R}^s)$ resp. $\in \mathcal{A}_q(\mathbb{R}^s)$, we consider the minimal class containing these which is invariant with respect to diffeomorphisms. Due to the functorial property of $\bar\mu_\varepsilon$, this class precisely consists of all images of constant test objects (in the sense just described) under the mappings $\tilde\varphi \mapsto ((\varepsilon, x) \mapsto \mathrm{pr}_1 \bar\mu_\varepsilon(\tilde\varphi, \mu^{-1} x))$ where μ ranges over all diffeomorphisms onto the open set under consideration. As the following example shows, the class of test objects obtained in this way (starting with all $\tilde\varphi \in \mathcal{A}_0(\mathbb{R}^s)$) is in fact different in general from $\mathcal{C}_b^\infty(I \times \Omega, \mathcal{A}_0(\mathbb{R}^s))$.

EXAMPLE 18.1. Let $\Omega := \mathbb{R}$; choose $\psi \in \mathcal{A}_0(\mathbb{R})$ satisfying $\psi(0) \ne 0$, $\mathrm{supp}\,\psi \subseteq [-1, +1]$. Setting $\phi(\varepsilon, x)(\xi) := \psi(\xi + \sin x)$ in fact defines an element of $\mathcal{C}_b^\infty(I \times \mathbb{R}, \mathcal{A}_0(\mathbb{R}))$. Now assume that there exists $\tilde\varphi \in \mathcal{A}_0(\mathbb{R})$ and a diffeomorphism $\mu : \tilde\Omega \to \mathbb{R}$ (where $\tilde\Omega \subseteq \mathbb{R}$ is open) such that $(\phi(\varepsilon, x), x) = \bar\mu_\varepsilon(\tilde\varphi, \mu^{-1} x)$ for all $x \in \mathbb{R}$. Setting $\xi := 0$, $x_1 := 0$, $x_2 := \frac{\pi}{2}$, respectively, we obtain $\psi(0) = \tilde\varphi(0) \cdot |(\mu^{-1})'(\mu(0))|$ resp. $\psi(1) = \tilde\varphi(0) \cdot |(\mu^{-1})'(\mu(\frac{\pi}{2}))|$. The first of these relations entails $\tilde\varphi(0) \ne 0$ while the second one implies $\tilde\varphi(0) = 0$, so we arrive at a contradiction.

In some situations, it may not even be desirable to require invariance of a Colombeau algebra with respect to all diffeomorphisms; for example, invariance only with respect to members of the Poincaré group might be of interest in applications in special relativity. This approach also could be combined with restricting the class of test objects to images of constant test objects under the particular group of transformations at hand. This opens the way to new classes of Colombeau algebras possessing weaker invariance properties than $\mathcal{G}^d(\Omega)$ does. However, these new objects still can be constructed on the basis of the scheme outlined in chapter 3 which, in our view, constitutes an appropriate general framework for the treatment of (full) Colombeau algebras.

Acknowledgments

The work on this series of papers was initiated during a visit of the authors at the department of mathematics of the university of Novi Sad in July 1998. We would like to thank the faculty and staff there, in particular Stevan Pilipović and his group for many helpful discussions and for their warm hospitality. In the course of a visit to Prague in the fall of 1999 we had the opportunity to meet J. Jelínek whose comments led to a number of corrections and improvements in the manuscript. Moreover, from his insider's point of view he introduced us to the prehistory of diffeomorphism invariant Colombeau algebras dating back to the late 1980s. We would like to thank G. Hörmann and M. Oberguggenberger for numerous stimulating discussions that importantly contributed to the content of the present memoir. Finally, we are indebted to Andreas Kriegl for sharing his expertise on infinite dimensional calculus.

Bibliography

[1] H. Balasin, *Colombeau's generalized functions on arbitrary manifolds*, gr-qc Preprint[1] **9610017** (1996).

[2] H. Balasin, *Distributional aspects of general relativity: The expample of the energy-momentum tensor of the extended Kerr-geometry*, in M. Grosser, G. Hörmann, M. Kunzinger, M. Oberguggenberger (Eds.), Nonlinear Theory of Generalized Functions, Chapman & Hall/CRC Res. Notes Math. **401**, Chapman & Hall/CRC, Boca Raton, FL, 1999, pp. 275–290.

[3] H. A. Biagioni, *A Nonlinear Theory of Generalized Functions*, (2nd ed.) Lecture Notes in Math. **1421**, Springer, New York, 1990.

[4] H. A. Biagioni, M. Oberguggenberger, *Generalized solutions to the Korteweg-de Vries and the regularized long-wave equations*, SIAM J. Math. Anal. **23** (1992), 923–940.

[5] H. A. Biagioni, M. Oberguggenberger, *Generalized solutions to Burgers' equation*, J. Differential Equations **97** (1992), 263–287.

[6] F. Berger, J. F. Colombeau, *Numerical solutions of one-pressure models in multifluid flows*, SIAM J. Numer. Anal. **32** (1995), 1139–1154.

[7] F. Berger, J. F. Colombeau, M. Moussaoui, *Solutions mesures de Dirac de systemes de lois de conservation et applications numeriques*, C. R. Acad. Sci. Paris Sér. I Math. **316** (1993), 989–994.

[8] J. F. Colombeau, *Differential Calculus and Holomorphy, Real and Complex Analysis in Locally Convex Spaces*, North Holland, Amsterdam, 1982.

[9] J. F. Colombeau, *New Generalized Functions and Multiplication of Distributions*, North Holland, Amsterdam, 1984.

[10] J. F. Colombeau, *Elementary Introduction to New Generalized Functions*, North Holland, Amsterdam, 1985.

[11] J. F. Colombeau, *Multiplication of distributions*, Bull. Amer. Math. Soc. (N.S.) **23** (1990), 251–268.

[12] J. F. Colombeau, *Multiplication of Distributions. A Tool in Mathematics, Numerical Engineering and Theoretical Physics*, Lecture Notes in Math. **1532**, Springer, New York, 1992.

[13] J. F. Colombeau, A. Meril, *Generalized functions and multiplication of distributions on C^∞ manifolds*, J. Math. Anal. Appl. **186** (1994), 357–364.

[14] J. F. Colombeau, A. Heibig, M. Oberguggenberger, *Le probleme de Cauchy dans un espace de fonctions generalisees I*, C. R. Acad. Sci. Paris Sér. I Math. **317** (1993), 851–855.

[15] J. F. Colombeau, A. Heibig, M. Oberguggenberger, *Le probleme de Cauchy dans un espace de fonctions generalisees II*, C. R. Acad. Sci. Paris Sér. I Math. **319** (1994), 1179–1183.

[16] J. F. Colombeau, M. Oberguggenberger, *On a hyperbolic system with a compatible quadratic term: Generalized solutions, delta waves, and multiplication of distributions*, Comm. Partial Differential Equations **15** (1990), 905–938.

[17] J. W. de Roever, M. Damsma, *Colombeau algebras on a C^∞-manifold*, Indag. Math. (N.S.) **2** (1991), 341–358.

[18] N. Dapić, S. Pilipović, *Microlocal analysis of Colombeau's generalized functions on a manifold*, Indag. Math. (N.S.) **7** (1996), 293–309.

[19] N. Dapić, S. Pilipović, D. Scarpalézos, *Microlocal analysis of Colombeau's generalized functions—Propagation of singularities*, J. Anal. Math. **75** (1998), 51–66.

[20] J. Dieudonné, *Éléments d'Analyse*, Vol **3**, Gauthier-Villars, Paris, 1974.

---

[1]available electronically at http://xxx.lanl.gov/archive/gr-qc

[21] E. Farkas, *Approximation properties of convenient vector spaces,* Preprint, Wien, 1996. [1]
[22] A. Frölicher, A. Kriegl, *Linear Spaces and Differentiation Theory,* Wiley, Chichester, 1988.
[23] H. Grosse, M. Oberguggenberger, I. T. Todorov, *Generalized functions for quantum fields obeying quadratic exchange relations,* Proc. Steklov Inst. Math. **228** (2000), 81-91.
[24] M. Grosser, M. Kunzinger, R. Steinbauer, J. Vickers, *A global theory of nonlinear generalized functions,* Preprint[2], 1999.
[25] L. Hörmander, *The Analysis of Linear Partial Differential Operators I,* Grundlehren Math. Wiss. **256**, Berlin 1990.
[26] J. Jelínek, *An intrinsic definition of the Colombeau generalized functions,* Comment. Math. Univ. Carolin. **40** (1999), 71–95.
[27] B. L. Keyfitz, H. C. Kranzer, *Spaces of weighted measures for conservation laws with singular shock solutions,* J. Differential Equations **118** (1995), 420–451.
[28] A. Kriegl, P. W. Michor, *The Convenient Setting of Global Analysis,* Math. Surveys Monogr. **53**, Amer. Math. Soc., Providence, RI, 1997.
[29] M. Kunzinger, *Lie Transformation Groups in Colombeau Algebras,* doctoral thesis, University of Vienna, 1996.
[30] M. Kunzinger, R. Steinbauer, *A rigorous solution concept for geodesic and geodesic deviation equations in impulsive gravitational waves,* J. Math. Phys. **40** (1999), 1479–1489.
[31] E. Landau, *Einige Ungleichungen für zweimal differentiierbare Funktionen,* Proc. London Math. Soc. Ser. 2, **13** (1913–1914), 43–49.
[32] M. Nedelkov, S. Pilipović, D. Scarpalézos, *The Linear Theory of Colombeau Generalized Functions,* Pitman Res. Notes Math. Ser. **385**, Longman, Harlow, 1998.
[33] M. Oberguggenberger, *Multiplication of Distributions and Applications to Partial Differential Equations,* Pitman Res. Notes Math. Ser. **259**, Longman, Harlow, 1992.
[34] M. Oberguggenberger, F. Russo, *Nonlinear SPDEs, Colombeau solutions and pathwise limits,* in L. Decreusefond, J. Gjerde, B. Øksendal, A. S. Üstünel (Eds.), Stochastic Analysis and Related Topics VI., Birkhäuser, Boston, 1998, pp. 319–332.
[35] E. E. Rosinger, *Distributions and Nonlinear Partial Differential Equations,* Lecture Notes Math. **684**, Springer, New York, 1978.
[36] E. E. Rosinger, *Non-Linear Partial Differential Equations. An Algebraic View of Generalized Solutions,* North Holland, Amsterdam, 1990.
[37] H. H. Schaefer, *Topological Vector Spaces* (5th ed.), Grad. Texts in Math., Springer, 1986.
[38] L. Schwartz, *Sur l'impossibilite de la multiplication des distributions,* C. R. Acad. Sci. Paris Sér. I Math. **239** (1954), 847–848.
[39] R. Steinbauer *Distributional Methods in General Relativity,* doctoral thesis, University of Vienna, 2000.
[40] J. A. Vickers, J. P. Wilson, *Invariance of the distributional curvature of the cone under smooth diffeomorphisms,* Classical Quantum Gravity **16** (1999), 579–588.
[41] J. A. Vickers, J. P. Wilson, *A nonlinear theory of tensor distributions,* ESI Preprint[3] **566** (1998).
[42] J. A. Vickers, *Nonlinear generalised functions in general relativity,* in M. Grosser, G. Hörmann, M. Kunzinger, M. Oberguggenberger (Eds.), Nonlinear Theory of Generalized Functions, Chapman & Hall/CRC Res. Notes Math. **401**, Chapman & Hall/CRC, Boca Raton, FL, 1999, pp. 275–290.
[43] J. P. Wilson, *Distributional curvature of time dependent cosmic strings,* Classical Quantum Gravity **14** (1997), 2485–2498.
[44] S.Yamamuro, *Differential Calculus in Topological Linear Spaces,* Lecture Notes in Math. **374**, Springer, New York, 1974.

---

[1] available electronically at http://diana.mat.univie.ac.at/~diana/dianapub.html
[2] available electronically at http://arxiv.org/abs/math.FA/9912216
[3] available electronically at http://www.esi.ac.at/ESI-Preprints.html

Editorial Information

To be published in the *Memoirs*, a paper must be correct, new, nontrivial, and significant. Further, it must be well written and of interest to a substantial number of mathematicians. Piecemeal results, such as an inconclusive step toward an unproved major theorem or a minor variation on a known result, are in general not acceptable for publication. Papers appearing in *Memoirs* are generally longer than those appearing in *Transactions*, which shares the same editorial committee.

As of May 31, 2001, the backlog for this journal was approximately 7 volumes. This estimate is the result of dividing the number of manuscripts for this journal in the Providence office that have not yet gone to the printer on the above date by the average number of monographs per volume over the previous twelve months, reduced by the number of volumes published in four months (the time necessary for preparing a volume for the printer). (There are 6 volumes per year, each containing at least 4 numbers.)

A Consent to Publish and Copyright Agreement is required before a paper will be published in the *Memoirs*. After a paper is accepted for publication, the Providence office will send a Consent to Publish and Copyright Agreement to all authors of the paper. By submitting a paper to the *Memoirs*, authors certify that the results have not been submitted to nor are they under consideration for publication by another journal, conference proceedings, or similar publication.

Information for Authors

Memoirs are printed from camera copy fully prepared by the author. This means that the finished book will look exactly like the copy submitted.

The paper must contain a *descriptive title* and an *abstract* that summarizes the article in language suitable for workers in the general field (algebra, analysis, etc.). The *descriptive title* should be short, but informative; useless or vague phrases such as "some remarks about" or "concerning" should be avoided. The *abstract* should be at least one complete sentence, and at most 300 words. Included with the footnotes to the paper should be the 2000 *Mathematics Subject Classification* representing the primary and secondary subjects of the article. The classifications are accessible from www.ams.org/msc/. The list of classifications is also available in print starting with the 1999 annual index of *Mathematical Reviews*. The Mathematics Subject Classification footnote may be followed by a list of *key words and phrases* describing the subject matter of the article and taken from it. Journal abbreviations used in bibliographies are listed in the latest *Mathematical Reviews* annual index. The series abbreviations are also accessible from www.ams.org/publications/. To help in preparing and verifying references, the AMS offers MR Lookup, a Reference Tool for Linking, at www.ams.org/mrlookup/. When the manuscript is submitted, authors should supply the editor with electronic addresses if available. These will be printed after the postal address at the end of the article.

Electronically prepared manuscripts. The AMS encourages electronically prepared manuscripts, with a strong preference for $\mathcal{A}_{\mathcal{M}}\mathcal{S}$-LaTeX. To this end, the Society has prepared $\mathcal{A}_{\mathcal{M}}\mathcal{S}$-LaTeX author packages for each AMS publication. Author packages include instructions for preparing electronic manuscripts, the *AMS Author Handbook*, samples, and a style file that generates the particular design specifications of that publication series. Though $\mathcal{A}_{\mathcal{M}}\mathcal{S}$-LaTeX is the highly preferred format of TeX, author packages are also available in $\mathcal{A}_{\mathcal{M}}\mathcal{S}$-TeX.

Authors may retrieve an author package from e-MATH starting from `www.ams.org/tex/` or via FTP to `ftp.ams.org` (login as `anonymous`, enter username as password, and type `cd pub/author-info`). The *AMS Author Handbook* and the *Instruction Manual* are available in PDF format following the author packages link from `www.ams.org/tex/`. The author package can be obtained free of charge by sending email to `pub@ams.org` (Internet) or from the Publication Division, American Mathematical Society, P.O. Box 6248, Providence, RI 02940-6248. When requesting an author package, please specify $\mathcal{AMS}$-LaTeX or $\mathcal{AMS}$-TeX, Macintosh or IBM (3.5) format, and the publication in which your paper will appear. Please be sure to include your complete mailing address.

Sending electronic files. After acceptance, the source file(s) should be sent to the Providence office (this includes any TeX source file, any graphics files, and the DVI or PostScript file).

Before sending the source file, be sure you have proofread your paper carefully. The files you send must be the EXACT files used to generate the proof copy that was accepted for publication. For all publications, authors are required to send a printed copy of their paper, which exactly matches the copy approved for publication, along with any graphics that will appear in the paper.

TeX files may be submitted by email, FTP, or on diskette. The DVI file(s) and PostScript files should be submitted only by FTP or on diskette unless they are encoded properly to submit through email. (DVI files are binary and PostScript files tend to be very large.)

Electronically prepared manuscripts can be sent via email to `pub-submit@ams.org` (Internet). The subject line of the message should include the publication code to identify it as a Memoir. TeX source files, DVI files, and PostScript files can be transferred over the Internet by FTP to the Internet node `e-math.ams.org` (130.44.1.100).

Electronic graphics. Comprehensive instructions on preparing graphics are available at `www.ams.org/jourhtml/graphics.html`. A few of the major requirements are given here.

Submit files for graphics as EPS (Encapsulated PostScript) files. This includes graphics originated via a graphics application as well as scanned photographs or other computer-generated images. If this is not possible, TIFF files are acceptable as long as they can be opened in Adobe Photoshop or Illustrator. No matter what method was used to produce the graphic, it is necessary to provide a paper copy to the AMS.

Authors using graphics packages for the creation of electronic art should also avoid the use of any lines thinner than 0.5 points in width. Many graphics packages allow the user to specify a "hairline" for a very thin line. Hairlines often look acceptable when proofed on a typical laser printer. However, when produced on a high-resolution laser imagesetter, hairlines become nearly invisible and will be lost entirely in the final printing process.

Screens should be set to values between 15% and 85%. Screens which fall outside of this range are too light or too dark to print correctly. Variations of screens within a graphic should be no less than 10%.

Inquiries. Any inquiries concerning a paper that has been accepted for publication should be sent directly to the Electronic Prepress Department, American Mathematical Society, P. O. Box 6248, Providence, RI 02940-6248.

Editors

This journal is designed particularly for long research papers, normally at least 80 pages in length, and groups of cognate papers in pure and applied mathematics. Papers intended for publication in the *Memoirs* should be addressed to one of the following editors. In principle the Memoirs welcomes electronic submissions, and some of the editors, those whose names appear below with an asterisk (*), have indicated that they prefer them. However, editors reserve the right to request hard copies after papers have been submitted electronically. Authors are advised to make preliminary email inquiries to editors about whether they are likely to be able to handle submissions in a particular electronic form.

Algebra to CHARLES CURTIS, Department of Mathematics, University of Oregon, Eugene, OR 97403-1222 email: `cwc@darkwing.uoregon.edu`

Algebraic geometry and commutative algebra to LAWRENCE EIN, Department of Mathematics, University of Illinois, 851 S. Morgan (M/C 249), Chicago, IL 60607-7045; email: `ein@uic.edu`

Algebraic topology and cohomology of groups to STEWART PRIDDY, Department of Mathematics, Northwestern University, 2033 Sheridan Road, Evanston, IL 60208-2730; email: `priddy@math.nwu.edu`

Combinatorics and Lie theory to SERGEY FOMIN, Department of Mathematics, University of Michigan, Ann Arbor, Michigan 48109-1109; email: `fomin@math.lsa.umich.edu`

Complex analysis and complex geometry to DUONG H. PHONG, Department of Mathematics, Columbia University, 2990 Broadway, New York, NY 10027-0029; email: `phong@math.columbia.edu`

*__Differential geometry and global analysis__ to LISA C. JEFFREY, Department of Mathematics, University of Toronto, 100 St. George St., Toronto, ON Canada M5S 3G3; email: `jeffrey@math.toronto.edu`

*__Dynamical systems and ergodic theory__ to ROBERT F. WILLIAMS, Department of Mathematics, University of Texas, Austin, Texas 78712-1082; email: `bob@math.utexas.edu`

Functional analysis and operator algebras to BRUCE E. BLACKADAR, Department of Mathematics, University of Nevada, Reno, NV 89557; email: `bruceb@math.unr.edu`

Geometric topology, knot theory and hyperbolic geometry to ABIGAIL A. THOMPSON, Department of Mathematics, University of California, Davis, Davis, CA 95616-5224; email: `thompson@math.ucdavis.edu`

Harmonic analysis, representation theory, and Lie theory to ROBERT J. STANTON, Department of Mathematics, The Ohio State University, 231 West 18th Avenue, Columbus, OH 43210-1174; email: `stanton@math.ohio-state.edu`

*__Logic__ to THEODORE SLAMAN, Department of Mathematics, University of California, Berkeley, CA 94720-3840; email: `slaman@math.berkeley.edu`

Number theory to MICHAEL J. LARSEN, Department of Mathematics, Indiana University, Bloomington, IN 47405; email: `larsen@math.indiana.edu`

*__Ordinary differential equations, partial differential equations, and applied mathematics__ to PETER W. BATES, Department of Mathematics, Brigham Young University, 292 TMCB, Provo, UT 84602-1001; email: `peter@math.byu.edu`

*__Partial differential equations and applied mathematics__ to BARBARA LEE KEYFITZ, Department of Mathematics, University of Houston, 4800 Calhoun Road, Houston, TX 77204-3476; email: `keyfitz@uh.edu`

*__Probability and statistics__ to KRZYSZTOF BURDZY, Department of Mathematics, University of Washington, Box 354350, Seattle, Washington 98195-4350; email: `burdzy@math.washington.edu`

*__Real and harmonic analysis and geometric partial differential equations__ to WILLIAM BECKNER, Department of Mathematics, University of Texas, Austin, TX 78712-1082; email: `beckner@math.utexas.edu`

All other communications to the editors should be addressed to the Managing Editor, WILLIAM BECKNER, Department of Mathematics, University of Texas, Austin, TX 78712-1082; email: `beckner@math.utexas.edu`.

Selected Titles in This Series

(*Continued from the front of this publication*)

699 **Alexander Fel′shtyn,** Dynamical zeta functions, Nielsen theory and Reidemeister torsion, 2000
698 **Andrew R. Kustin,** Complexes associated to two vectors and a rectangular matrix, 2000
697 **Deguang Han and David R. Larson,** Frames, bases and group representations, 2000
696 **Donald J. Estep, Mats G. Larson, and Roy D. Williams,** Estimating the error of numerical solutions of systems of reaction-diffusion equations, 2000
695 **Vitaly Bergelson and Randall McCutcheon,** An ergodic IP polynomial Szemerédi theorem, 2000
694 **Alberto Bressan, Graziano Crasta, and Benedetto Piccoli,** Well-posedness of the Cauchy problem for $n \times n$ systems of conservation laws, 2000
693 **Doug Pickrell,** Invariant measures for unitary groups associated to Kac-Moody Lie algebras, 2000
692 **Mara D. Neusel,** Inverse invariant theory and Steenrod operations, 2000
691 **Bruce Hughes and Stratos Prassidis,** Control and relaxation over the circle, 2000
690 **Robert Rumely, Chi Fong Lau, and Robert Varley,** Existence of the sectional capacity, 2000
689 **M. A. Dickmann and F. Miraglia,** Special groups: Boolean-theoretic methods in the theory of quadratic forms, 2000
688 **Piotr Hajłasz and Pekka Koskela,** Sobolev met Poincaré, 2000
687 **Guy David and Stephen Semmes,** Uniform rectifiability and quasiminimizing sets of arbitrary codimension, 2000
686 **L. Gaunce Lewis, Jr.,** Splitting theorems for certain equivariant spectra, 2000
685 **Jean-Luc Joly, Guy Metivier, and Jeffrey Rauch,** Caustics for dissipative semilinear oscillations, 2000
684 **Harvey I. Blau, Bangteng Xu, Z. Arad, E. Fisman, V. Miloslavsky, and M. Muzychuk,** Homogeneous integral table algebras of degree three: A trilogy, 2000
683 **Serge Bouc,** Non-additive exact functors and tensor induction for Mackey functors, 2000
682 **Martin Majewski,** ational homotopical models and uniqueness, 2000
681 **David P. Blecher, Paul S. Muhly, and Vern I. Paulsen,** Categories of operator modules (Morita equivalence and projective modules, 2000
680 **Joachim Zacharias,** Continuous tensor products and Arveson's spectral C^*-algebras, 2000
679 **Y. A. Abramovich and A. K. Kitover,** Inverses of disjointness preserving operators, 2000
678 **Wilhelm Stannat,** The theory of generalized Dirichlet forms and its applications in analysis and stochastics, 1999
677 **Volodymyr V. Lyubashenko,** Squared Hopf algebras, 1999
676 **S. Strelitz,** Asymptotics for solutions of linear differential equations having turning points with applications, 1999
675 **Michael B. Marcus and Jay Rosen,** Renormalized self-intersection local times and Wick power chaos processes, 1999
674 **R. Lawther and D. M. Testerman,** A_1 subgroups of exceptional algebraic groups, 1999
673 **John Lott,** Diffeomorphisms and noncommutative analytic torsion, 1999
672 **Yael Karshon,** Periodic Hamiltonian flows on four dimensional manifolds, 1999
671 **Andrzej Rosłanowski and Saharon Shelah,** Norms on possibilities I: Forcing with trees and creatures, 1999

For a complete list of titles in this series, visit the
AMS Bookstore at **www.ams.org/bookstore/**.